The Urbana Free Library

To renew materials call
217-367-4057

6-08

	DATE DUE	
DEC 13 2008		
DEC 1 7 2010		

NIGHT FIRE

RONNIE GREENE

NIGHT FIRE

Big Oil, Poison Air,
and Margie Richard's Fight
to Save Her Town

Amistad
An Imprint of HarperCollinsPublishers

HarperCollins books may be purchased for educational, business, or sales promotional use. For information please write: Special Markets Department, HarperCollins Publishers, 10 East 53rd Street, New York, NY 10022.

FIRST EDITION

Designed by Janet M. Evans

Library of Congress Cataloging-in-Publication Data

Greene, Ronnie.
    Night fire : big oil, poison air, and Margie Richard's fight to save her town / by Ronnie Greene. — 1st ed.
        p. cm.
        ISBN 978-0-06-112362-7
        1. Petroleum industry and trade—Social aspects—Louisiana—Norco. 2. Shell Chemical Company. 3. Pollution—Louisiana—Norco. 4. African-American neighborhoods—Louisiana—Norco—Environmental conditions. 5. Richard, Margie Eugene. 6. Environmental health—Louisiana—Norco —Citizen participation. I. Title. II. Title: Big oil, poison air, and Margie Richard's fight to save her town.

HD9568.N67G74 2008
363.738'70976333—dc22   2007042653

---

08 09 10 11 12 OV/RRD 10 9 8 7 6 5 4 3 2 1

TO LINDA GREENE,
BETH REINHARD,
ABBY AND EMMA GREENE

Three generations, four lights

CONTENTS

NIGHT FIRE

Part
ONE

DEATH
IN
DIAMOND

It was summer 1973 and the moist bayou air clung to the homes and the people of southern Louisiana. Gibson's, the discount shop, offered relief: water sprinklers for $2.47, electric fans for $8.88, coolers for $1.47. The school calendar was about to turn, so any parent could tell you on-the-cheap specials filled the aisles: girls' blouses for $2.67, boys' khakis for $4.77, pleated skirts just $3.22. The August Super Savings Sale filled a full page of the weekly *L'Observateur*. Long-timers didn't need directions to the cut-rate hub of choice—just hop on Airline Highway, pass the swampy pea green waters on your right and the billowing chemical plants and oil refineries on your left, until you rolled right into LaPlace, pronounced "LaPlaz," a shopping emporium for the small towns dotting this piece of Louisiana low country, some thirty short minutes and a full world away from big city New Orleans.

The Sugar Queen of St. John the Baptist Parish was newly minted—Miss Stephanie Ann Delaneuville—but the night of glittering smiles and cabaret show tunes masked the difficulties creeping into this pocket of Cajun country. On Central Avenue in Reserve, the same town that produced the baby-faced queen, a forty-eight-year-old woman was stabbed twenty-two times and strangled with a television cord in her husband's auto store. Police said two teen brothers committed the grisly murder. In neighboring St. Charles Parish, a deputy died from a bullet fired by his own revolver after he turned out to quell a disturbance at a nightclub along River Road. Five people attacked him, and one fired the officer's .357 Magnum into his back. The twenty-four-year-old deputy took his last breath on the operating table.

In Louisiana, the seat of government is a parish, not a county, and the parishes of St. Charles, St. John the Baptist, and St. James lock arms like brothers, nestled between New Orleans to the east and Baton Rouge up Interstate 10. Their towns don't grace the pages of shiny tourist brochures, but the oil industry knows the address.

In southern Louisiana, crude rules. Industry didn't mischievously creep into the region, like the nasty crime wave; it loudly and openly overtook it. Two industrial facilities fill every one mile of asphalt between New Orleans and Baton Rouge—more than 150 plants in an eighty-mile stretch—and most of them hover over poor or middle-class neighborhoods. Drive along River Road or Airline Highway at night, and the evening sky alights like a dazzling cityscape. Pull closer, and it's not some bustling downtown district, but a stretch of industry locals call the Chemical Corridor, or Cancer Alley, where oil refineries and chemical plants churn smoky flares to the skies, towering over the tiny clapboard homes beneath them. On clear days the distinctive white smoke puffs into its very own cloud, announcing the industry's foothold miles in advance. Chemical Corridor is so powerful, locals will tell you, that it makes its own weather.

In the summer of 1973, the newly formed Energy Corporation of Louisiana Ltd. announced it would build a $300 million refinery on the grounds of a former plantation in nearby Garyville, in St. John the Baptist Parish, with promises of churning out 200,000 barrels of petroleum a day. Near Convent, in St. James, Texaco was expanding its River Road refinery to produce 200,000 barrels in a day—on top of the 145,000 already coming from its site next door. The Texaco plant stood just off the Sunshine Bridge along the Mississippi River; the roadway itself was erected during the administration of Louisiana's own singing governor, Jimmie Davis, most famous for his ditty "You Are My Sunshine."

In Norco in St. Charles Parish, Shell, the biggest player of them all, was finalizing plans for the latest expansion of its chemical plant. Built in the 1950s and not done growing, the Shell facility towered, quite literally, over the four-street neighborhood known as Diamond.

The Shell Chemical Company plant was the giant's second major industrial unit in Norco, along with the refinery across town. In 1916, the New Orleans Refining Company bought 366 acres of sugarcane field from the Good Hope Plantation, and a decade later, in 1929, the Shell Petroleum Corporation purchased the refinery. Norco took its name from the New Orleans Refining Company, and of course the label fit. Practically the entire town was sandwiched between the refinery on the east end, 15536 River Road, and the newer chemical plant on the west, 16122 River Road.

The chemical plant's growth spurt would ultimately stretch it to nine football fields in length. Its aboveground pipelines moved a toxic mix of propane, ethylene, propylene, and natural gas, yet its security fence hovered little more than twenty feet from the nearest row of Diamond properties. Shell portrayed its plant as a pristine symbol of progress, with jobs for the community and goods for consumers. Ethylene becomes antifreeze, brake fluid, and detergents; propylene turns into plastic milk bottles, paints, and auto parts. Butadiene, yet another chemical mixed on-site, helps create nylon carpet, automobile tires, and latex.

For two decades, Diamond families tried to maintain a peaceful coexistence with their neighbor, yet by 1973 the constant foul odors and growing number of illnesses prompted mounting complaints. The community's young children suffered from asthma, and the elderly were afflicted with bronchitis, pneumonia, and even cancer. No one at Shell or in the parish government paid them much heed. After each complaint, the company replied firmly that all was well, and the plant continued to churn, its pipes hissing and belching at all hours, sending smoke plumes into the air and forcing residents to shutter their homes to avoid the sour smell of chemical burn-offs. The industry calls it flaring, and it's meant to burn off noxious gases from inside the plant during production breakdowns. Diamond residents witnessed the intense fire and smoke produced during flaring and wondered just what chemicals might be seeping into their homesteads, and what damage those chemicals might be wreaking upon their families.

Diamond's small, clapboard ranch houses, some tidy, some not, dotted Washington, Cathy, Diamond, and East streets like rows in a cornfield. Norco was a mostly white town of 3,500, but the Diamond district was all black, a 269-lot community relegated to the flood-prone banks of the Mississippi. Dead-end streets and a block-thick grove of trees separated Diamond from its more well-to-do neighbors.

In white Norco, Shell was the center of pride, the bread-winner that put food on the table and sometimes sent kids off to college. People from Diamond landed few of Shell's high-paying jobs, and the community's streets were graced with no fringe benefits and just two exits out, a practical problem in the case of emergency and a symbolic one for hardscrabble homeowners who tended the land with their hands.

Diamond homes stood on soil that once housed a sugar-cane plantation, and on January 8, 1811, the largest slave revolt in U.S. history made a stop on this very ground. Several hundred insurgent slaves armed with cane knives and clubs marched down River Road toward New Orleans, killing two whites, burning crops, and gathering weaponry along the journey. The revolt was brought to a bloody halt in nearby Destrehan by a militia and U.S. troops, forces that combined to kill sixty-six slaves in battle. Sixteen more were sentenced to death thereafter, the final justice coming in crude fashion—they were decapitated, and their heads were placed atop poles along River Road as a warning. Historians say seventeen other slaves remained unaccounted for, bringing the documented toll of dead or missing to ninety-nine. Diamond residents could trace their family histories and find an ancestor among the slaves who worked this land, the community's history of labor and bloodshed binding them together.

In this tiny community, neighbors were more than friends; they were often family. Step inside any Diamond homestead, and you'd likely hear of an aunt, a cousin, a brother, or an in-law who lived just a few doors down, or maybe two streets over. A family tree, spread over four small streets.

Twenty-six Washington Street stood center stage in the community, its green and white décor glistening like a giant welcome sign. Neighbors and family popped by for Theodore Eugene's prized okra or butterbeans, sold by the chicken coop out back. Theodore, "Uncle Brother" they called him, pocketed what you could pay: the vegetables in his garden carried no price tag and sometimes went for no cost at all; his cherished eggs were the only item that came with a charge affixed. Other residents shuffled in for sage advice from this wiry man with roots in the community as solid as a neighborhood pecan tree. Then there was the cooking of Mabel Eugene, "Aunt Mabel" they called her, whose portions were hearty enough to share. Mabel cooked in a restaurant in the white part of town, so it was natural that she offered up the best jambalaya, fried okra, and mustard greens at her own dining room table. Mabel sang in the church choir, penned inspired church speeches, and talked, talked, talked—no surprise, perhaps, for a child born into a family of sixteen children, several of them still residing in Diamond.

Brother stood medium tall and lean, with round glasses and a serious countenance neighbors knew meant trust when they dropped in for counsel on their taxes or their children's schooling. He always wore a blue striped suit when talking business, using a voice so low it sometimes sounded like a whisper. In Diamond, that whisper was gospel.

"Brother, when you going to finish messing around in that yard and get something done?" Mabel shouted from the window.

"Mabel, when you going to *do something?*" Husband came back.

"I'm already finished. Just waiting on you."

And they shared a laugh, and went about their business.

Opposites in so many ways, they were bonded on February 16, 1936, the day that Theodore crafted a love poem to his future bride while studying at Xavier University. "Darling, I am still thinking of you—night and day. You just seem to stay on my mind. You couldn't imagine how much I pray for June to come so I can see you." Theodore's spare words carried so much deep meaning that years later one of Mabel's grand-daughters preserved the hand-scrawled note inside a glass frame as a keepsake. Theodore and Mabel wed in St. Charles Parish on January 26, 1939. One year later their first child, Naomi, arrived, and in December of 1941 came Margie, each girl entering the world in Charity Hospital in New Orleans.

At first the family lived in a shotgun shack not far from Diamond, in the neighborhood of Belltown—"shotgun shack" because you could open the front door, fire a shotgun, and the bullet would shoot straight out the back.

Brother was among the first black police deputies on the local force, and his maiden arrest was of a neighborhood boy who had been sassing his mom so much the police were dispatched to shush this disturbance of the peace. Theodore didn't cuff the child; he sat him down and spoke to him like a son. "You come from a good family. You don't need to do this," he told the boy.

Brother believed in rehabilitation, and he'd sometimes hire ex-cons to do odd jobs around the yard.

"Jailbird, jailbird," Margie and Naomi sang to one on a broiling afternoon.

"Come here to me," came Daddy's voice. "I don't ever want to hear you two call people jailbird or making fun of people. No matter what happens in life, people deserve a chance. I don't want to see you teasing." He grounded the girls that day, banishing them to the front porch.

Another day Margie found one dollar outside and hopped back into the house with her sudden treasure, her eyes dancing. Daddy heard her story of discovered gold and stopped her right there. "If you found it, child, it's not yours. If you find something, that means somebody lost it," he said, sending Margie out to find the rightful owner. She knocked door-to-door until she did.

Uncle Brother read the encyclopedia and could tell you just about anything you wanted to know about the world and its exotic locales. Anything, that is, except how to get there. He rarely, if ever, left the small town of Norco. Mabel didn't either, and her rare ventures out always required someone else's assistance. Though she excelled as a backseat driver, she never did learn how to drive a car.

Theodore Eugene ran with a group of old-timers who called themselves the Oddfellas, and they came from Norco, Montz, New Sarpy, St. Rose, and Good Hope, all little-known outposts filled with squat one-story homes in the shadow of industry.

Black high-school students living in St. Charles Parish had

no school of their own. Instead, they had to catch a bus to attend school in another city, Kenner, a growing, mostly white suburb of New Orleans one parish over, in Jefferson. "When my girls go to high school, they're not going to have to catch a bus to the other side of town," Theodore, the Oddfellas' leader, declared when his daughters were in grade school, and Bethune High School would rise within Norco's limits well in time for Margie and Naomi to join the class roster.

Theodore was secretary of the group that worked to force the school board's hand, and at times the battle became so intense that each side thought only a fistfight would settle it. Cooler heads prevailed, and Uncle Brother played a role in keeping the calm. One day, with the fight still engaged, the white superintendent asked a group of black residents who their official spokesman was. The group didn't know it needed an "official spokesman," but after the question was posed to them, all eyes turned to Uncle Brother—he would be their leader. He worked with his tight band of friends, going to court to secure an order ensuring that the black section of St. Charles Parish would have a high school of its own.

When Bethune High opened in 1952, it had two typewriters, compared with the white high school in Destrehan with forty. Still, the battle was considered won. "What we wanted was high schools for our black children, and we got it," Theodore said. Mary McLeod Bethune, the showpiece's namesake, spoke on its ground not long before her death, and Margie Eugene felt the words wash over her and settle in.

"Never start out to lose the race."

In 1953, not long after the opening of Bethune, Shell seized the plots of land that housed the Eugenes and many other

Belltown families, to build its chemical plant. Forced to move, many residents packed their belongings and traveled less than one mile to the Diamond neighborhood.

The 1950s were a time of progress in southern Louisiana, and not just for Shell. St. John Parish issued $1.8 million in bonds to build plants to provide water and natural gas for the entire community. Lutcher Motors looked to tap into this spending spree by showcasing its new model: *Save with a '54 Studebaker.* Maurin's theater was showing movies every Sunday at 3:30, leaving plenty of time for church. There was a new Sugar Queen for the state of Louisiana, and she was local—Miss Rita Fay Coco, of Reserve, and such a fetching sight that they flew her to New York City to celebrate. Not coincidentally, it seemed, the local sugar harvest was up, and with the weather just right, strawberries were in such bloom that local growers needed four hundred new pickers. Fill a six-pint carrier, pocket twenty cents; do nine pints, get thirty.

Meantime, the powerful oil industry braced for a bare-knuckle fight. With a state gas tax increase looming, the petroleum giants devised a plan of attack, gathering at the Club Café in Reserve with words so tough you'd think they were filing off to war. Industry leaders declared that higher gas taxes would derail the region's economy, leaving them no choice but to launch a lobbying and public relations campaign "to defend oil men and the consumers of oil products from . . . the present burden of taxes," *L'Observateur* reported. "Parish Oil Men Plan Tax Fight" said the headline, and there was no mistaking the force they'd muster to protect their spoils.

The Eugenes had their own, more modest vision of prosperity. In 1954, Theodore paid $360 cash for the double lot on

Washington Street that had space for a home on one side and vacant land on the other, right across from the Shell chemical plant, and there the family settled. He was now working in construction, and his job soon shifted to Waco, Texas, but Mabel wasn't ready to leave Louisiana. So Theodore took early retirement and the family stayed in Diamond, his hobbies—his crops, his pigs, and his chicken coop—becoming his vocation. Margie was twelve, and Naomi fourteen.

In dollars and cents, many, though not all, of the families in Diamond were poor. But the roots were rich, and on Mother's Day or Thanksgiving or Christmas, the grills and stoves burned late and the mingling stretched on past midnight and into a new day, the chatter rising to the chorus of a family get-together.

Twenty-six Washington Street might as well have been called the Diamond Community Hall, as it served as the gathering place for neighborhood activists to plan upcoming church events or hash out concerns over jobs and schooling. Just 1,270 square feet, the home seemed much bigger, with a living room, dining room, and kitchen area all open to one another, so a neighbor could sit in one room and quite naturally visit with Brother or Mabel in the other.

Step inside and the first thing you'd notice was the scent of Mabel Eugene's signature vegetable soup simmering on the stove. Peer out the back kitchen window and there'd be Theodore pulling his turnips and carrots from the ground.

From the Eugene family's front yard now, you don't have to look far to see the Shell chemical plant—so close that an errant ball tossed by a child could clear its razor-wire security fence. Attached to Shell's security fence, and facing Diamond, are a half dozen signs warning residents of the chemicals being handled on-site: "Warning NGL Pipeline," "Ethylene Pipeline," "Chlorine Pipeline," "Natural Gasoline Pipeline," "Propane Pipeline," and finally, "Warning Propylene Pipeline."

Margie and Naomi were raised here, and knew the home's heartbeat well. Margie was the adventurous, wide-eyed child who had gone duck hunting with Dad at the age of eight and later ran the fifty-yard dash for the high-school track team, while Naomi took on Theodore's stoicism and studied home economics. Naomi was like Dad, but she was the mama's girl. Margie was like Mom, and you knew she was Daddy's girl.

Margie was outspoken and always insisted on seeing things with her own two eyes. As a child, she'd pull apart the doll she was given for Christmas just to see how it had been made. Sister would help her Scotch-tape it back together so Mom and Dad wouldn't know. Playing school as girls in the room they shared, Naomi stood up front, ruler in hand, pointing at an imaginary chalkboard. One day she gave in and handed the chalk to younger sister. "Time for recess," declared Margie.

Though Norco seemed like two towns, Margie and Naomi absorbed their parents' preaching: Don't stereotype others or let others stereotype you. Not even when the local theater forced the girls to sit upstairs, separate from the whites, and "Colored Only" signs peppered their youthful travels.

Theodore saw something in the fast feet and quick mind of his youngest, the girl who talked in the animated manner of a performer. Margie used her hands to tell stories, and her eyes shone as she came to the conclusion of a particular anecdote about friends and family, who were one and the same.

Theodore listened to his girl go on and on once again,

filling their home with grand stories and big dreams, and finally he hushed her and turned to his brother, George Eugene. "That girl's gonna be a big traveler in life," he said, and now they all were smiling. Theodore always insisted that his girls read the paper and learn about the world, and he envisioned his own children visiting those faraway places he had never seen.

As she peered out her window from 26 Washington, Margie's dreams raced beyond Diamond, a community where a ticket to college was by no means a sure thing, and where more than a few children had stopped their schooling well before high-school graduation to work or raise families of their own. After graduating from Bethune High, Margie Eugene went on to study at Grambling College, where she earned a bachelor's degree in 1965. Sister Naomi graduated from Grambling too, no surprise, and the Eugene women became schoolteachers. Margie focused on health, physical education, and social studies, and Naomi applied her knowledge of home economics and early childhood development.

Now adults, the sisters were together at 26 Washington Street on a steamy Thursday August afternoon in 1973. Margie, thirty-one, a recently divorced single mother with two kids at home and many more in her classes at New Sarpy Middle School, had swung by to visit her sister, mom, and dad. Naomi and her husband were living at 26 Washington, and she was pregnant with her second child. Theodore had added a third bedroom for his kin.

When the explosion came, it shook the house to its foundation.

The oven door blew wide open, spattering sizzling grease

from a baking ham onto Naomi's legs, leaving what would become a lifelong scar. Margie, the high-school track star, sprinted out the front door and into the street. Her neighbors spilled from their tiny homes in scattered frenzy.

They converged at lot 22, square 14, the Washington Street home of Helen Washington. Mrs. Helen wasn't much of a talker, but neighbors savored her coconut and sweet potato pies, and children admired the seventy-two-year-old woman's long, beautiful hair. Her late husband had made the bricks for their house by hand, a grinding labor neighbors had witnessed with quiet admiration. Mrs. Helen had shot into the air like a cannonball with the violent blast. Now she lay motionless on the pavement, while her house burned to the ground. Margie inhaled the bitter smell of Mrs. Helen's singed hair.

In a flash, Leroy Jones, a respectful sixteen-year-old who did chores around the neighborhood, emerged from the rubble, fleeing down the street as if he could outrun the flames that engulfed him. Moments earlier the boy had been talking with neighborhood pals. Some smelled a strange odor in the air and biked on home. Leroy had a job to finish, mowing Mrs. Helen's lawn.

The explosion lit into his mower like a torch. Emergency crews descended on Diamond and airlifted the boy to the hospital, but death took him several days later.

Margie's aunt Inez, the community seamstress, and cousin Doris Pollard lived next to Mrs. Helen, and it was as if the fire that torched Mrs. Helen's house was marching straight to Inez's with more killing in mind. Inez Dewey weighed nearly four hundred pounds and was bound to a wheelchair, and when Doris heard the giant boom, she peered out the window and witnessed the family tree afire. The flames raced straight from Mrs. Helen's house toward them, blockading the front door, so Doris frantically wheeled her mother out the back, as far as they could go. She spotted her own five children in the small family swimming pool, their faces frozen with confusion. "Go on over to Cathy Street!" Doris shouted. "Just get out of the pool and go to the back over the fence and go to the next road. Go!"

She turned to her mother, a woman so immense that lifting her over the back chain-link fence to safety was not possible. "Mama, don't worry, because I'm not going to leave you."

Sobbing now, her eyes filled with a mix of panic and love, mother turned to child. "You have your life to live. You have your children to think about."

Doris had already sent her children to safety. She kept her hand on Mama's wheelchair.

Albert Ducree, a neighbor with a Vietnam battle scar on his leg, scanned the confusion and glimpsed the women huddled together, caged in by the fence. He hopped it, rolled Inez down the ramp, and, with every ounce of might, lifted her to safety, but as he reached down to raise her, his already wounded leg caught on the fence, and the blood began to gush. He saved Inez, but Albert was the one now rushing to the hospital.

Margie watched the man lift beloved Auntie 'Nez to safety, and her mind raced. She thought of her lifeless neighbor and the burning boy running for rescue. She inhaled smoke, smelled death, and the discomfort was so raw she turned her gaze.

As neighbors scattered and debris cascaded in the air, the image of the stark sheet over Mrs. Helen's body burned a lasting memory in Margie's mind.

Slowly, Margie's view shifted toward the hulking chemical plant standing directly across from Mrs. Helen's now leveled home, Inez's home, her family's home, all of them on Washington Street. When Shell's chemical plant had opened two decades earlier, the company had promised good jobs for its neighbors. People in Diamond could tell you that was a hollow promise, as far as they were concerned. The good Shell jobs had been given to homesteaders over in white Norco, not here. Here, in black Norco, were smoke, fire, and chaos.

The disaster was caused by a cluster of Shell pipelines that had caught fire, ruptured, and exploded. "There was a pipeline and there was a leak. The young man who was cutting the grass ignited it," a company official said, explaining the accident.

In the days to follow, Margie waited for the company to aid the Diamond community in its time of crisis, to show respect for the dead, to counsel the wounded. She kept looking for Shell, but the visits did not come.

Outside Diamond, the two deaths barely elicited a ripple. After just a couple of days, the news was buried in a few paragraphs in the local newspaper, and then it went away for good.

Margie sang in the church choir at the time, and at Mrs. Helen's funeral, tears flowed from her eyes. She looked around and saw her neighbors, the small community mourning together. Oh God, she thought. I don't see one representative of the plant at this funeral. Not even one crisis team.

Mrs. Helen's property was left that day in "ruins and rubble," as if an earthquake had come to town and singled it out for destruction. Shell later bought the vacant lot, a symbol of that deadly summer day of 1973, for $3,000.

At Leroy Jones's funeral, the world's second-largest oil company quietly handed his mother $500, a token gesture that would remain secret for more than twenty-five years.

SISTERS
AND
SECRETS

Naomi Eugene Sterling was vacuuming the floor that afternoon when she felt the familiar pain down in her chest. She shut off the vacuum, thinking another good cough would clear her lungs. Her Kleenex turned red.

Since her midtwenties, Naomi's health had grown worse. Her breath became shallow, forcing her to stop short until normal breathing returned. Her body grew tired, sending her to the couch at home or in the teachers' lounge at school. It was as if she were constantly allergic to one thing or another, and her sneezing spells ushered in the seasons. Through it all, Naomi never made a fuss. She'd show up at the doctor's when her body ached beyond tolerance and undergo some tests, but most times she'd simply grab an extra dose of allergy medicine and be on her way. In a day, maybe two, she'd feel refreshed.

Naomi didn't schedule time in the day for fretting. She had readings to teach the elementary-school kids, a husband to tend to, and three children to raise.

And, of course, she had Margie.

The sisters were so different on the surface—quick-tongued Margie, reserved Naomi—but in truth they were inseparable, and as children they shared a single bed. Of the Diamond neighborhood's four streets, Washington stood closest to the plant, yet growing up in its shadow did not stop Margie and Naomi from having a normal childhood. Naomi's hobby was dressing hair, and she'd curl Margie's just right. Margie was the makeup specialist, and she'd touch up Naomi's just so. The girls played hopscotch and jumped rope out front, except when they played school.

Margie grew up quickly, and Naomi was the lone soul who could share her sister's secrets. Margie excelled at school and sports, but she was a bit of a rebel in the straitlaced Eugene family—always on the go, quick with her words, while Naomi hovered around Mama's skirt, a homemaker already as a teen. "Hey, Mom and Dad," Margie once asked her parents, "am I adopted?"

In her early years in high school, Margie made close acquaintance with a neighborhood boy everyone called Pete. In an era when "sex education" was just beginning to reach public schools and teenagers having sex was strictly taboo, she became pregnant.

Margie kept her secret for months. She wore extrawide skirts to school, and she avoided her parents' close gaze, but she did confide in Naomi.

"You need to tell Mom and Dad," Sister told her, straight to the point, as always.

"Will you be with me when I tell them?"

"You need to do this on your own."

Margie was frozen with fear, dreading that her disgrace would bring shame on the family, and that her parents would force her to marry the boy. Tossing one night in bed, she could hear her parents talking in the other room. Theodore and Mabel were no dummies. They had figured out Margie's secret. Straining now to listen, Margie heard her father's soft voice.

"We've got to support her," he said. Hearing the words, Margie settled in her bed and fell asleep at last.

The next day Daddy talked to Margie. "We're going to take care of the baby, but both of you are going to get your education," Theodore Eugene said.

Margie worried about kids teasing at school. "You cannot listen to everything you hear," her father said.

In 1958, as a sixteen-year-old high-school sophomore, Margie Eugene gave birth to a baby girl, Caprice. She finished her high-school studies, as Mabel and Theodore tended to their grandchild while Margie took classes at Bethune High.

Naomi, the oldest, soon went off to college, and one day a letter arrived in the mail for Margie. It was from Naomi, telling her sister to stop being so hard on herself.

As Margie prepared for college, the family decided that Caprice would live with an aunt in California so Margie could finish her studies. Both Margie's and Pete's families adored the new baby, and Margie told herself that with so much affection blanketing her, Caprice had been born into good fortune.

Yet with her girl seemingly a world away in California, Margie ached. "Dad, I'm going to stop college because I want to be home with my baby."

Theodore replied, "You can make it. That's my Margie, and I know you can finish it."

She graduated from Grambling in 1965, and a year later, she and Pete were married. In 1968, the couple had their second daughter, Ericker. Pete had served in the army, and by then was doing bookkeeping work by day and taking business classes at night.

Later he became close with one of Margie's fellow church choir members. The marriage fell apart, and Margie was so hurt by the affair that she changed churches. The high-school sweethearts divorced in 1972. Pete married the choir singer, and they had their own children, and though the divorce devastated Margie, she remained close with her ex's family.

With Margie at a painfully low point, Naomi was the one who sat Sister down for a talk. "Life offers challenges, Sister, but look at all you've got."

Margie had always set out to succeed: to run fastest in track, to roast the tastiest pig with Dad at the neighborhood fair, to be the best teacher in the entire school. Now she confronted a failed marriage in a family where "till death do us part" had real meaning, and Sister's words took time to settle in. The soothing helped, but Margie's ailing would not be cured completely. One day she had showed up at her daughter Caprice's school to sign the child's report card, and her oldest girl looked at her. "Mother," she asked, "why do we have to have different names?"

"If I never remarry, out of respect for the children, I will just use Richard," Margie told herself, keeping her ex-husband's last name.

Thirty-something, suddenly single Margie Eugene Richard plunked a Wellington single-wide trailer on Mom and Dad's property across from the Shell plant in 1975 and moved from the nearby town of Montz, population 1,100. Although just two years earlier, an explosion at the plant had killed Leroy Jones and the elderly Helen Washington, moving back home came naturally. She didn't have to pay rent on Dad's land, which meant something to a single mom rearing two kids on a teacher's pay. On the same plot of land Daddy had bought twenty-one years earlier, Margie made a home at 28 Washington Street, in 924 square feet of living space, with three bedrooms, and walls covered with knickknacks and family glossies. Mom and Dad knew to leave their back door unlocked so Margie and her girls could stop in at all hours. Margie came to 26 Washington Street so often that you could follow the well-worn path from back door to back door.

Margie and the girls managed their finances as best they could. Almost every day Margie or one of her daughters would fill an old mason jar with spare change. When bills were due and money was thin, Margie'd holler out, "Time to open the emergency jar," and the girls poured the bounty out onto the table, the change clanging like found treasure. Margie took an occasional odd job at Shell, moonlighting a few hours when the company needed extra hands filling and measuring chemicals, a job they called the "turnaround." Margie helped supervise the measuring but stopped after discovering that she had been handling dangerous toxins. She also sold Mary Kay products, perfecting her technique on seventeen-year-old Caprice and seven-year-old Ericker, who pronounced her own name *Erica*.

When southern Louisiana's prosperity abruptly ended, every dollar mattered. Margie and her girls settled into the trailer not long after the U.S. gas shortage of 1974 had pummeled a region brimming with oil refineries but now seemingly bereft of fuel. Things got so out of hand that a doctor felt compelled to pen a newspaper editorial with a shocking warning: "Gasoline siphoners risk quick death." With the public desperate for gas and looking for a target on which to vent its fury, the refineries took the heat, and Shell hurriedly booked full-page ads in *L'Observateur* defending itself. "How in all conscience can anyone call these excess profits?" one ad declared, accompanied by a lengthy statement from the president of the Shell Oil Company. A month later, Shell placed another advertisement—"Why you can't get all the gasoline you want"—explaining that government rules allowed farmers, emergency vehicles, factories, businesses, utilities, and airlines all to stand in line before consumers. "Now you know why some Shell stations are sometimes forced to hang up a sign reading 'Sorry, No Gas,'" it said.

Racial tensions roiled southern Louisiana too, and they sprang from one of the places the Eugene sisters held dearest, the public schools. The region was locked in a bitter fight over school desegregation, and nowhere was the conflict more profound than at Destrehan High, less than four miles from Norco. By this time, Bethune High School had closed its doors, so students from St. Charles Parish's black neighborhoods attended schools that were predominantly white.

The fighting shook Caprice, now a Destrehan High freshman. As Margie had finished studies at Grambling College, Caprice had grown up in San Francisco in the early 1960s, where she lived with an aunt and witnessed Haight-Ashbury and the beginnings of the hippie movement and people of diverse skin colors intermingling without fuss or fury. When Caprice came home after Margie's schooling was finished, she was a child with a worldview greater than that of Norco, Louisiana. Black children weren't supposed to act differently than kids with white skin, or vice versa, and Caprice settled in with relative ease as a rare black child in the nearly all-white Sacred Heart of Jesus Elementary School. The Catholic school stood on Spruce Street in white Norco, not even one mile from the family's Washington Street home. Yet with Diamond full of dead-end streets, Margie had to take a series of extra turns to reach the school, adding nearly a mile to the drive. She'd start by heading in the opposite direction to River Road, the main thoroughfare for the community, and then travel a half mile to Apple Street on the other side of town. She'd turn left on Apple for sixth-tenths of a mile, then make another left on Fourth Street for four-tenths of a mile, and then finally a right on Spruce. For Margie, Caprice's journey to school was a vivid reminder that in the event of a catastrophe, Diamond residents had few easy ways out of their neighborhood. The community was nearly boxed in.

Racial divides still existed, even for children. Caprice had been in the Brownies in San Francisco, yet when she went to join a chapter in her hometown, the door was shut. Margie went to make things right, and finally she did, convincing the local chapter that Caprice Richard could join the Brownies like any other child. By the time Mom got the ruling overturned, Caprice the tomboy was on to other hobbies, no longer excited enough to join. When a local nun said Caprice couldn't join the softball team, Margie paid another visit and again got the rule changed. This time Caprice followed through. In high school, Caprice witnessed Friday-night football games become violent not only on the field but off it, with fistfights between black and white students.

One October day in 1974, not long after a football-night scuffle carried over into the school halls, the atmosphere became so tense that the school announced it would shut down early. Teachers hustled twelve hundred students onto school buses, rushing them away from the fighting and yelling. A group of whites surrounded the buses carrying black students, firing bottles, rocks, and soda cans, the teens' faces turning beet red as they shouted profanities.

Caprice was sitting on a school bus when she heard a gunshot pierce the air, and the shouting die with it. Timothy Weber, a thirteen-year-old white child waiting for his mother away from the fracas, fell to the ground, and hours later he was dead. Witnesses fingered the shooter as someone with a gloved hand. Gary Tyler, a sixteen-year-old black youth, appeared to fit the description.

The killing brought more rage, and the Ku Klux Klan patrolled Destrehan's streets, drawing state police, armed in riot gear, to stand a shaky watch. "We will help protect white people from rampaging black savages and murderers!" a young Klan leader named David Duke announced.

The Black Panthers came too, from the Ninth Ward of nearby New Orleans.

An all-white jury convicted Tyler of the killing and sentenced him to death, ignoring pleas from the black community that the boy was innocent, and making the sixteen-year-old the youngest inmate on death row, until the United States Supreme Court later struck down the state's death penalty law and handed him life behind bars.

Caprice had attended elementary school with the dead boy's sister, and Margie had taught the child in middle school. She had also taught members of Tyler's family.

A massive crowd turned out for Timothy Weber's funeral on October 10, 1974, the mourners filling every corner of Norco's Sacred Heart of Jesus Church. Into this anguished crowd entered Margie, Caprice, and Ericker Richard. Margie had heard the rumors that the KKK might be around that day, but she believed it her duty to pay respects to the boy and his family. She had taught him in school, after all, and Caprice was very close to his sister.

It could have been my child, she thought. I can imagine how that mother felt.

Yet, pulling up to the church, Margie was surprised not to see a single other black face. Other black teachers had taught Timothy Weber in school too. Where were they? Suddenly, she noticed people staring at her and her children as they walked inside the church.

When Margie and her girls walked out after the service, a reporter, spotting the only black faces among the mourners, rushed to them. "Are you related to the family?" the reporter asked. Taken aback by the query, Margie could only think to point out her own skin color to the reporter. No, she wasn't related; she'd taught the boy in school.

Just then the dead child's mother, Leah Weber, stepped out of the church and, in tears from the wrenching service, came to Margie. "Thank you so much for coming," she said.

The mother was enraged at the killer, but let Margie and her daughters know they were welcome. The feeling didn't extend to all of southern Louisiana. With the KKK on one side and the Black Panthers on the other, and media outside of Louisiana drawn to this racial fracture, blacks and whites kept separate quarters.

Margie had not planned to follow the procession to the cemetery, but after Timothy's mother embraced her, and after taking stock of the racial divide so evident even during the grieving over a child's death, she and her girls traveled to the Destrehan grave site. "Girls, let's say our prayers and we are going to support this family all the way," she said.

Again, a TV camera pointed toward them. This time Margie was not in a mood to chat. Ericker, a curious child of six, turned toward the camera, and the TV news that day broadcast a snippet of the family. Around town, Margie heard a taunt whispered from some quarters. "Honky lover."

After the services, Margie talked with her father.

"I went out of respect, so it didn't matter to me what people said," she told him, but then admitted: "If the tension was that high, Dad, maybe I shouldn't have gone."

"You did the right thing," Theodore said.

Inside her own middle-school classes, Margie now preached that justice should be color-blind, and she had no better example to make her case than the injustice in Destrehan. The death of a child and the fighting between whites and blacks filling her with sadness, anger, and fear, Margie spoke to Caprice one day not long after the shooting. "I can't rest, Caprice, with you being in high school there. We've made a decision, but it won't be forever."

Margie sent her teenage child back to California to stay with relatives.

Caprice spent a full year away from home, keeping good grades in Los Angeles, until the fighting and fear subsided enough for Margie to arrange for her to come back home, just as the family was settling into its new Diamond trailer in 1975.

Sister Naomi settled too. Her family had lived on Washington Street for years, and Naomi envisioned having her own homestead in Diamond. She owned a plot of land in the community, but the 1973 explosion frightened Naomi, who was pregnant at the time, and she soon sold the empty lot to Shell. When her new home was ready in nearby Kenner, the Sterling family moved in.

One afternoon in 1978, with Naomi's Kleenex changing hue, Margie's phone rang. It was Naomi, calling from Charity Hospital.

"Don't panic, Margie. I'll be fine." Sister's voice was calm, naturally. Then she broke the news: "I coughed up blood today and the doctors said I have tuberculosis."

Margie cried out to God, then said, "Naomi, I'll be there."

Hanging up the phone in stunned silence, Margie prayed silently until a gentle rap on her trailer's back door shook her from her private pain. An old friend from the area, the Reverend Curtis Stacey, was in Diamond visiting his mother and had decided to pay Margie a quick visit.

"Naomi's sick," she told the reverend, gathering her things and rushing off to Charity. Rev. Stacey followed, and together they said a prayer of faith in divine healing inside the same hospital that had welcomed both girls to the Louisiana soil nearly four decades before.

Looking at her sister hooked to machines in the hospital bed, Margie thought of Ericker. Years earlier, Margie's youngest had grabbed her chest and fallen to the ground while playing outside, unable to breathe. Margie hustled her into the back of her yellow Pinto and sped to the hospital, then listened as doctors explained how Ericker's lung had collapsed.

"All sickness is not death," Margie told herself as the medical team stuck her girl with a needle that seemed endlessly long. Seven days later, Ericker went home. The child still suffered from asthma and needed three shots a week, but she was on the mend. Caprice helped her younger sister through the discomfort and the shots. Margie watched as Caprice, exactly as Naomi had done for her, provided strength to Ericker. In truth, it was no surprise. Caprice was a mirror of Aunt Naomi and Grandpa Theodore. She got things done without much fanfare.

Margie looked back at Naomi on the hospital bed and knew it was her turn to be the pillar. Sister spent three long days at Charity, and Margie prayed she would find recovery, just as Ericker had.

As they prepared to release her, doctors delivered surprising news. A battery of tests showed that Naomi did not have TB after all. Margie and Mabel Eugene cried tears of joy, yet their sobs couldn't dispel the mystery that continued to hang over Naomi's sickness. Doctors couldn't pinpoint what caused her to cough blood in the first place, and they sent her home with medicine the family hoped would be the cure.

A year later, Naomi broke out sneezing on her way to school. This will pass, she thought, striding to the young faces in her classroom, and she managed to teach her lesson. When the bell rang, Naomi retired to the teachers' lounge, but the cigarette smoke was so suffocating that she had to escape outside, where the children lined up for recess. She sneezed again, this time so hard she could feel something pop inside. The blood was steady and constant. She collapsed on the asphalt.

Back at the hospital, the diagnosis now was that Naomi had sarcoidosis, a disease that causes inflammation of the body's tissues and often targets the lungs, where it produces small lumps. Doctors say environmental factors can trigger the disease, and Naomi's upbringing had been smothered in smoke and flares from the Chemical Corridor that was their home.

Although treatable in the early stages, the disease can scar sensitive tissues if left unchecked. Doctors prescribed the steroid prednisone, the drug most commonly used to combat the disease and among the most effective—unless administered *after* sarcoidosis has caused irreversible scarring. Naomi was diagnosed late in the disease's march through her lungs.

Weight gain, one of the drug's side effects, caused a once-thin Naomi to balloon to size 18. She turned so ill that doctors hooked her to a pure-oxygen machine to fend off infections that might aggravate her condition. During particularly bad spells, she couldn't be outside at all. Her water had to be purified, her air completely clear. Amid the jumble of steroids and machines, with one lung destroyed, she now lost weight and shrank to a size 8.

As Naomi deteriorated, Margie ached. She'd sometimes drive by Naomi's house and see everyone inside, but couldn't pull in. Naomi wouldn't stand for crying, and knowing she couldn't keep her composure, she'd drive off with eyes welling.

"Daddy, I can't take it," she said. Theodore Eugene gave her his strength, and the next day Margie pulled into the driveway. This time she did not cry.

Naomi shuffled between home and hospital so often that the family half joked that East Jefferson General Hospital, where she was receiving follow-up care, should name a wing after her. One evening she was back at Jefferson, and as Margie hovered over her, she felt Naomi slipping from her.

"Don't leave me," Margie said, their bodies so close their cheeks met. Sister opened her eyes in the hospital bed. "I'm not going anywhere." She survived another crisis, and Margie counted her blessings, until the next time Naomi was rushed back.

No longer fearful, Margie stopped in every day, and the sisters shared their secrets. One night the TV was on and a funeral scene filled the screen. Naomi smiled. "See there, they're at a funeral but they're not all dressed in black," she told Margie.

A few days later, Naomi looked over at Margie and told her to go home, fix up her hair, and get some rest. "Put the roses by my bed, and send Mama on in." Margie didn't want to leave and fussed, but Naomi reminded her who was oldest. By the time Margie got home she learned that Naomi Eugene Sterling had passed quietly, survived by husband Milton Sterling, three children, and hundreds more she had taught in nearly twenty years of school. Her death came twenty-three days before Christmas 1983.

Naomi's passing was officially declared on medical records as caused by the relatively mysterious lung disease sarcoidosis. She was forty-three.

At the funeral, Margie sang a heartfelt hymn and didn't break down, keeping a promise to herself and wearing, as most of the mourners that day, bright colors to celebrate life rather than death. When pallbearers carried Naomi away, her promise broke:

"How will I move on? What will I do?"

In her trailer across from the chemical plant, Margie turned to the Bible. She felt cold and alone, Naomi no longer there to provide shelter, and found Psalm 37. "The just will possess the land and live in it forever. The mouths of the just utter wisdom; their tongues speak what is right. God's teaching is in their hearts; their steps do not falter."

She placed the Bible atop her small coffee table and stepped into her front yard. There, staring her down as it had for much of her life, the Shell plant spewed its mist into the air, and the white noise of industrial machinery rang out inside her. She smelled the air and thought of her sister, of Mrs. Helen burned to death under the stark white sheet, and of Leroy Jones racing through the street in flames. Diamond residents, nearly every one of them, knew how the plant made them, their neighbors, and their children sick, and now Margie's sister had had her life cut short by too many decades to fathom. All the while, Shell never acknowledged fault or responsibility, save for the donation at Leroy Jones's funeral.

Three

PEARL HARBOR

Almost two years to the day after her sister's passing, Margie gained a gift that helped soften her burden. Grandson Christopher, the first child of Margie's oldest daughter, Caprice, was born December 4, 1985. The winter holidays had traditionally been an inspired time for the Eugene family before Naomi's passing, and Christopher's arrival was like a rebirth, especially for Margie.

Christmas night, all the family gathered at one house or another in Diamond, and everyone brought a special dish: a vat of gumbo, a bowl of mustard greens, a moist turkey, a goblet of turkey dressing. The hosts, often Aunt Mabel and Uncle Brother, baked a ham, and the eating and mingling carried a sweet chorus, the robust chatter sounding, after years lost, like a family reunion.

For Margie, Christmas could be Christmas again, even as

problems with the plant across the road continued as part of Diamond's daily fabric.

Several months before Christopher's birth, inspectors with the Louisiana Department of Environmental Quality had unearthed some troubling matters with Shell Chemical Company's management of hazardous waste. An inspection that April found that the company had skirted Louisiana's hazardous-waste regulations by failing to properly manage containers holding hazardous waste; by failing to conduct a weekly inspection of locations where the hazardous containers were stored, or to maintain proper records for their inspection; and, among other problems, by storing waste "in an environmentally unsound manner," the official state order said. The company was directed to immediately clean up its act—by properly managing the containers used to handle hazardous waste, improving its inspection and record-keeping procedures, and doing a better job of labeling the containers. By June of that year, Shell Chemical Company told the state it had implemented all the required corrections. The plant was operating just fine, Shell said, and the potentially harmful issue quickly faded, or so it seemed.

Little more than a year later, Shell Chemical skirted environmental-protection laws once again. This time it was fined $77,500 for the illegal release of 580,000 pounds of ethylene into the atmosphere. Exposure to ethylene, a highly flammable chemical and an explosive hazard, can cause headaches, dizziness, or even unconsciousness. Shell paid up, and the plant continued to churn, the minor citation or fine barely registering a ripple in a region clogged with petrochemical plants and refineries. Diamond residents most fearful of the plant had little avenue to voice their worries, as Shell operated in the good graces of the parish political and bureaucratic establishment. The company said it was operating cleanly and efficiently, the occasional state notation aside, and few in power would dare challenge that contention. The chemical plant went back to its business, and the residents of Diamond to theirs.

As the calendar turned and the years brought back some sense of normalcy, Margie resettled into the bustle of her life, with physical education and social studies classes for the schoolkids, Bible studies in her trailer, weekends with family, friends, and church. Margie coached track, volleyball, and softball at the school too; her full schedule helped keep at bay the sadness over Sister's death, at least for a few hours each day.

Naomi, the former home economics student, pretty with an angular face and sharpness in her eyes, had carried herself with a seriousness unmistakably like her father's. Her smile had been glowing, but one you earned. In the classroom, Naomi had been nurturing but no-nonsense, and the elementary-school kids had listened.

With her students, Margie was Margie. "GOD DON'T MAKE NO DUMMIES," she wrote in all capital letters on the chalkboard, finishing the thought with, "AND YOU'RE NO DUMMY." She ruffled some feathers at New Sarpy Middle for bringing God's name into the equation, but using a slang that spoke to her students grabbed hold of the young minds.

Now a single mother, Margie frequently turned to the Bible for support. She often found Psalm 37: "Do not be provoked by evildoers; do not envy those who do wrong. Like grass they wither quickly; like green plants they wilt away. Trust in the Lord and do good that you may dwell in the land and live secure."

In the classroom, Margie favored rehabilitation, not punishment, just as her father had when he employed ex-cons to work in his yard. She chaired local substance-abuse programs long before "Just say no" was fashionable. She organized a trip to Louisiana's Angola Prison for the troubled students, and excursions to Washington, D.C., for the well behaved. Some children couldn't afford the trip, so Margie set aside her personal stipend for the D.C. excursion, put the names of needy children in a hat, and pulled one out, which allowed that child to take a trip that would have been missed otherwise. When kids in her PE classes didn't wear socks, because their family couldn't afford them, Margie went into her daughters' bedroom drawers and pulled out an extra pair or two. One night a young girl ran away from home, and the police scoured St. Charles Parish, fearing the worst. Margie's phone rang at 3 A.M. and it was the girl, turning to Ms. Richard for help.

Get out of line with Margie or skip school, and she wouldn't send you to detention or suspension. No, better that you show up on a Saturday morning in Diamond, where Margie would set you to work cleaning yards of the community's elderly residents, or working at Uncle Brother's hog farm when the extra hands were needed. Aunt Mabel would counsel the young troublemakers, and then put out a plate of food for them. Margie's way was rarely by the book.

One afternoon school administrators assigned Lamar, a serial skipper, to her middle-school history and social enrichment classes. The child had been in practically every other teacher's class, but none could contain him, so Margie did a little homework of her own, dialing up the boy's family to discover that he enjoyed drawing and, of all things, flourished working in the garden. She learned this before Lamar set foot in her class.

"You're not going to skip my class, Lamar; you're going to stay right here," Teacher told the new boy in class.

"Ms. Richard, can I go to the bathroom?"

"No," came Teacher's reply, knowing that Lamar would use any ruse to skip class. You send him to the bathroom, you might as well give him a pass to hightail it to the playground.

"Can I go get my tablet?"

"No."

"Can I go get my pencil? I need it!"

"No. Look in your bottom drawer—there's a pencil and tablet there. You're going to stay in this class. I know you can do gardening well, you can plant things, and I know you can draw. You're not going to leave this class until you finish."

The boy was stunned that his teacher knew the first thing about him. Margie set his chair right next to hers, and she gave him his assignments: planting flowers around the flagpole outside school for his enrichment class, appeasing the gardener in him, and copying every picture in the history book, appeasing the artist in him.

The serial skipper took out his notebook and pencil, and day after day he sketched pictures from the history book, careful to mimic the photos and to include a few words about each. Weeks later, he finished the book, and by the way, he didn't skip class.

"You ever thought about going around the world speaking to people?" he asked Ms. Richard another day.

"What do you think I'm doing here, Lamar?"

"No, I mean helping people?"

"What do you think I'm doing here, Lamar?" Teacher said again. "Lamar, you can do just like I'm doing."

Soon, Margie's trailer filled to capacity. Chris was nearly two and a half when his mom and dad temporarily moved in with Margie. Caprice had twelve hours of course study to finish at Xavier University, where she was preparing for a career in science lab work, and her husband suggested that instead of taking out another student loan to pay for it, they live in Margie's trailer for a year to save money. Caprice; her husband, Allen Torregano, a burly former construction worker who had recently taken work as a policeman; and Chris stayed in the master bedroom at the back of the trailer. Ericker, Margie's now college-age daughter, bunked in a sleeping bag in bedroom number two. Margie slept in number one, closest to the front door.

At 3:30 A.M. on May 5, 1988, the trailer was silent, everyone asleep inside. Suddenly, Margie's full-size mattress catapulted off the bed frame with the force of a shotgun blast.

"The plant is on fire!" Margie wailed. "The plant is exploding!"

Ericker screamed, "An earthquake!"

Panicked, Margie scampered outside; her son-in-law, towering over her at six feet two and 260 pounds, was already there, equally panicked. The pitch-black sky sparkled with fire, turning night into day; the air was so thick with smoke it was nearly impossible to breathe. The cloud of fire was so immense that it looked as if the earth itself were exploding. Other families, wearing robes and pajamas, some in underwear and bare feet, scattered to their front porches and into the streets, and they witnessed a fiery ball of red and yellow blanket Norco like a mushroom cloud.

Pearl Harbor, Margie thought, jolted by the vision of Diamond in flames once more, before she realized that this time the source was the Shell refinery across town. Margie taught Pearl Harbor in her history class, and the powerful photographs captured on that infamous day had long ago made a strong impression on her. Now, in the middle of the night, she witnessed a roiling cloud of fire and smoke that would sear itself into her memory: Pearl Harbor had arrived in Norco.

The sixty-year-old Shell oil refinery that hugged the Mississippi stood barely one mile from the Shell chemical plant that anchored Margie's neighborhood. Shell bookended the town of Norco, with the chemical plant straddling Diamond on the west end, and the refinery, which now processed more than two hundred thousand barrels of crude each day, anchoring the east. Many refinery workers lived in Norco's white section, which stood closest to this Shell facility.

The plant's explosion arrived with no warning except for the unfortunate few close enough to feel its thunder building from inside Shell's sixteen-story catalytic cracking unit, where crude was refined into gasoline. Investigators would soon conclude that a corroded eight-inch vapor line was the trigger to the blast, causing an "instantaneous line failure" that released some seventeen thousand pounds of hydrocarbon vapor for thirty seconds. The unit's superheater furnace was the likely ignition source.

The explosion hit like a tornado and then swept over Norco with the force of a hurricane, collapsing ceilings, unhinging doors, and delivering tremors thirty-five miles away. Homes were leveled and storefronts damaged, the blast ripping through the Bank of St. Charles, pummeling the Ace Hardware Center, blowing the wall off the Norco co-op supermarket, and shattering the stained-glass windows of Sacred Heart of Jesus Church. Houses were crushed, mobile homes cracked, residents wounded by glass and flying debris. Cars were covered with Venetian blinds, ceilings came down, floors glittered with shards of glass. A BLESS THIS HOUSE sign hurtled into the air, tumbling unceremoniously to the ground. Norco residents, in bed at that hour, looked up to see their ceilings crack in half right above them.

"It went boom, and then boom again," one resident, sixty-five, told reporters on the scene. "It knocked me to the floor, and as I was getting up, it knocked me down again."

Inside the plant, the men closest to the blast had no escape. Seven Shell workers—men with a combined seventy years' experience—were killed. Some were so badly burned that forensic pathologists from the FBI and Louisiana State University had to use dental records to identify them. Killed in the plant were Ernie Carrillo, Bill Coles, Lloyd Gregoire, John Moisant, Jimmy Poche, Joey Poirrier, and Roland Satterlee. Five of these men had been inside the catalytic cracking unit's control room when the explosion ripped through the plant. A sixth died approximately thirty feet outside the control room, and the seventh was killed when the blast's negative pressure literally sucked a brick wall toward the control room, where it engulfed him.

Nineteen other Shell employees, working in different units of the refinery at the time, suffered injuries—eighteen men and one woman, ranging in age from twenty-seven to fifty-seven. Six were hurt badly enough to be hospitalized, and the rest were treated on-site for cuts, contusions, burns, and fragments of shrapnel that lodged in their eyes. Yet rescue crews had to wait hours, until the fire faded and the wreckage cooled, before they could enter the giant refinery.

Shell managers, awakened in the middle of the night with word of the explosion, likewise could not make their way to the refinery until many hours later. They would not learn of the full death toll until the company compared its roster of employees with the names of those missing, and until the forensic experts definitively identified the dead.

Outside the plant, more than twenty residents were injured, some pinned under fallen debris, some scraped by shattered glass.

When Margie first felt the fire and smoke cascading toward Diamond, she hustled her family into her car and headed to nearby Montz, finding refuge in a relative's home.

At first Theodore and Mabel Eugene would not go. When Margie's household came to usher them out, Theodore was firm. He had not heard the official Shell whistle go off ordering everyone to leave, and without hearing that alarm, he would stay right where he was.

Mabel wouldn't budge either. "I've been married to him for fifty years. If he won't leave, I am going to stay with him," she said.

Margie and her son-in-law pressed the matter, and finally Mama and Daddy relented; but Theodore Eugene's mind would not give him rest, and Uncle Brother spent many evenings after the explosion walking up and down the street, buried in thought.

Margie's mind churned too. "The whole town panicked. I just knew it was going to be a spontaneous combustion. My mobile home had a crack, straight down a vertical line. My mom and dad's house shifted on its sills. Buildings were collapsed all on Apple Street in the white section."

Margie's wasn't the only family to have fled in the middle of the night. On Cathy Street in Diamond, Gaynel Johnson, a nursing assistant with three children ages twelve to sixteen, was used to being up in the middle of the night. For a spell, she had worked the "dog shift" at the nursing home from eleven at night until seven in the morning. Though she had recently attained the more comfortable three-to-eleven P.M. shift, Gaynel was so accustomed to being awake while everyone else slept that she was on her knees saying a prayer when, abruptly, all she could see was dust in the hallway. The entire ceiling had come down.

"Wake up!" she shouted. "We've had an explosion."

Gaynel hollered some more, but her husband, a laborer, didn't awaken, so she shook him until he did. "What?" Eddie Johnson Sr. said, foggy from sleep.

"Let's go!" Gaynel shouted with her smoky, dead-serious voice, hurriedly helping her children get their clothes on. Just then she witnessed her husband scamper to the front steps, a panic-stricken man in his underwear, with one hand on his heart and the other on his head.

In white Norco, Sal Digirolamo, a three-decade veteran of the Shell chemical plant who was recuperating from a hernia operation a few days before, was sound asleep when the explosion occurred. It wouldn't be long before Sal was back on the job, as Shell was practically part of the family. Two of his sons worked for Shell, one at the chemical plant and the other at the refinery, and his father and no fewer than five brothers-in-law had likewise been employed by Shell's refinery. Sal is built like solid oak, and his voice is pure southern Louisianan, rich and expressive, with a cadence unique to the region: not exactly Cajun, but more in tune with the dialect of St. Bernard's Parish near New Orleans. "Margie" in Sal's voice is "Maw-gie." There's no mistaking his loyalty to Shell and his hometown, as anyone on the Norco Civic Association can attest.

Sal had always been a hard sleeper, so he missed the loud and eerie hissing noise, which broke his wife's sleep. She shook him from bed. Sal walked to the kitchen and saw a window blown out. Instantly he thought of his son, and the company man panicked. Was Joey on shift that night? Sal struggled to think and realized, thankfully, that his son was not working at the refinery.

Though his house on Oak Street suffered relatively little damage, when Sal walked out his front door, he began to grasp just how extensive the explosion had been. Sal looked to the plant, still ablaze at that hour, and knew tragedy had come to Shell, but it would be hours before he knew just how serious.

Wreckage filled the streets of both black and white Norco. Hundreds of millions of dollars in damage had been wrought, but the destruction was so severe that you couldn't put a price tag on it, at least not yet. Theodore and Mabel Eugene needed a new roof. Margie needed new flooring. Practically everyone needed something, and it would be months before houses were homes again.

With smoke still choking the air, investigators from Shell and the U.S. Occupational Safety and Health Administration swooped in to determine a cause. Political leaders and government officials descended on the plant and the community, and with a parade of media before them, they assured Norco's residents they would be made whole. OSHA cited several serious violations of workplace safety laws, and Shell vowed to make improvements and get the plant back up to full speed as soon as possible. The company held a memorial on the refinery grounds for the fallen workers, and years later it would dedicate a monument to honor the men.

Margie watched the media come in, saw Shell send its emissaries, and heard the politicians promise justice. Her thoughts turned back to 1973, when the blast had come from her side of Norco. Where had the cavalry been then?

The 1973 explosion had attracted little attention, but this one had far-reaching implications. Authorities had ordered thousands to evacuate, including everyone in Margie's four-street Diamond neighborhood and thousands in the white section by the refinery. Then there was the death toll. Yet, from the Diamond community's viewpoint, the powers that be seemed focused on the impact beyond Norco. Choking off Shell's gushing force of petroleum would affect the entire nation, as consumers from California to Maine would be forced to reach a little deeper to fill up their tanks. The loss of this single refinery, even for the short term, meant the depletion of nearly 1 percent of the nation's gas supply, and on the New York Mercantile Exchange, unleaded gasoline gained almost a penny a gallon in value. With the world's gaze on Norco, government officials repeatedly promised to prevent tragedy from reoccurring, and Shell had already raised the topic of monetary payouts and internal improvements.

As the familiar plants in Margie's backyard flared anew, a truth took hold of her. It wasn't that her Diamond neighbors had been dealt a poor hand, she realized. Rather, they had never been given a hand at all. When industry staked its claim to Norco, it arrived promising a goody basket full of jobs and fine living for the entire community. Yet now, three full decades since Shell Chemical settled in across from Diamond, those promises echoed hollowly, as empty as the vacant lots that had begun cropping up in some patches of the community. Politicians and Shell officials spoke often of the economic jewels industry provided, but all the people of Diamond knew was their four-street neighborhood sandwiched between a chemical plant and a refinery.

In the weeks immediately following the catastrophe, Margie and her neighbors began sleeping fully clothed, ready to flee should another explosion jolt them from bed. They missed work, lost sleep, and visited many a doctor.

Unable to rest one sticky fall night, Margie Richard reached for her Bible once more. What she envisioned was inspired by the presence of her first grandchild, Christopher Torregano. He needed to grow up where he could play outside in fresh air and live the life of a normal child. Margie thought of his future, and she knew that the days of the community merely griping about the Shell plant had to end. Raised never to complain just for the sake of complaining, Margie believed it was best to do something meaningful or just leave well enough alone.

Feeling conviction, not anger, she readied to place herself at the forefront of a movement that would consume her personal and professional life for the next fourteen years.

Four

A MISSION

A year after the catastrophic explosion, Diamond residents gathered one evening inside the quaint Masonic meeting hall on Washington Street. The hall looked every bit like someone's small wood frame house, but in Diamond, it served as the place to meet for official business or social gatherings. The sessions were typically filled with chatter over the winter weather and school schedules and the crops out back. Local college students had helped residents spiff it up, scrubbing the interior and splashing a fresh coat of paint on the exterior. The facelift was fitting, for tonight the expressions inside the hall were stern, and what the community envisioned was another facelift, though on a larger scale—for the entire Diamond community.

Like a broken record, residents spoke of the smoke, the deaths, and the worries over sick children and battered homes.

Even Margie had become susceptible to anxiety attacks. When workers turned on a lawn mower outside her window at school, she jumped in the air and shrieked. She began missing school days, one after the other, and turned to therapists and church leaders for counseling. At night, the time she felt most uneasy, she began taking medication to help her fall asleep.

Janice Darensbourg, a Diamond resident suffering from bronchitis and sinus troubles, had invited a longtime family friend, attorney Allen J. Myles, to the hall that night. Myles listened politely as residents spoke of life near the plant, and then delivered a beam of hope. He was working with another so-called fenceline community in Louisiana, with the goal of securing a relocation settlement that would allow the impoverished residents to leave big industry's shadow. The strategy was to convince industry of the harm of having people so close to its plants, prompting the companies to buy out neighboring properties at a price high enough to allow residents to afford homes elsewhere. Diamond could do this too, Myles told those gathered around him, but first it had to organize by setting up a grassroots committee and picking a leader.

A few names bounded around the hall. Sitting up front, Janice knew that none rang true. She thought of her poor health and of how her eight-year-old son Larry was already a veteran at the local hospital, battling never-ending bouts with asthma, and she flashed back to the 1988 explosion that shattered her Sheetrock and windows, and to the 1973 blast that left her in constant fear. Janice told herself that what Diamond needed was someone on the ball enough to know facts from fibs and packing enough gumption to look Shell in the eye.

"Can I nominate someone?"

The crowd turned to her. "I nominate Margie Richard," Janice said. "She's a good leader, a church person. She's outspoken. She's educated."

Margie wasn't even in the building at the onset of the meeting. She happened to be out of town, and when she got home she asked her mother what was going on and headed over to the Masonic hall, arriving just as the meeting was coming to a close.

When she arrived, the lawyer was preparing to pass out retainer agreements for residents to sign, the first step in what would evolve into a long, contentious process to grab their fair share from Shell.

Margie absorbed the bustle around her, and as she entered the room, her neighbors had a message for her.

"We elected you," Janice told her. Janice was Margie's longtime friend from Belltown, where the families had lived in squat shotgun shacks until Shell bought the land, and the people moved to Diamond. "It's time."

For Margie, it was like a confirmation, and she had a message of her own.

"If I am going to lead we need to get organized."

"Okay," Janice said. "We can meet at my house next week."

Several nights later, the interested community members gathered at Janice's, and the lawyer came back to town.

"If we're going to do this, let's do this right. Let's elect a president," Margie said.

"You're the president," a resident shot out, and this time it was official.

The words froze Margie for a flickering moment. Then she scanned the room and saw her friends gathered there, many of them elderly, and she thought of how sickness had invaded all of their families.

"Oh Lord." Margie smiled. "If ya'll picked me as the leader, and you think enough of me to do it, then yes, I'll accept it."

For years Margie had been sharing her concerns with anyone who listened, and the seed for her leadership of Diamond had been planted long ago—after the 1973 blast, after her sister's 1983 funeral, and now, again, after the 1988 explosion. Her mission was to join with her neighbors and press a multinational behemoth to move them away from the menacing flares.

Years ago Margie's grass had died, and she wondered if the plant's air was to blame. She called Shell, and the company dispatched someone to her trailer to gather a sample of the parched grass. Weeks later Margie's phone rang, and it was Shell. The air was just fine, the company said, and the chemical plant had nothing to do with her decaying yard. Still one who insisted on seeing things with her own eyes, Margie had decided she'd learn for herself if that spoiled grass could be attributed to the Shell plant. And now that she'd been anointed Diamond's leader, she would apply a more fervent determination to find out if Shell caused her neighbors' sickness as well. "It's time," she said. This was no screaming gadfly Diamond long-timers turned to, but a woman who heard her father preach equality for all, and who believed words without acts meant nothing.

"My ancestors actually came over here chained and bound. I admired the bravery in the conditions they had. I wasn't chained. I wasn't bound. I suffered from a different type of slavery, but I knew I was free because Dad taught us you are somebody because you are made in God," she explained. "I began to look deep into my inner spirit. 'Oh God, if this is what you called me to do . . .' Knowledge that is not applied, to me, is not worth anything. I have to apply what I am learning."

Gaynel Johnson, a Cathy Street resident, was in the crowd those evenings. The mother of an eleven-month-old child during the 1973 explosion, she was deeply rattled by the catastrophe. Her health had always been poor, but the 1988 blast, which jolted her from her bed, drove home the point that she and the rest of the community needed relief. Many times after the Shell plant flared, Gaynel rushed to the hospital. She suffered from bronchitis and walking pneumonia, and since the explosion of 1988, she had seen three psychiatrists to control the flashbacks, insomnia, and worry. At night she'd awake choking from the chemical scent, and in the morning the air drove her from her porch. During one visit to Charity, the hospital where she was born in November 1950, the doctor looked her over, scanned her test results, and posed a question.

"Did you ever work at a chemical plant?"

"No," Gaynel replied brusquely. "I never worked at one, but I live like a sandwich between two."

Gaynel, if you don't get out of there, you are going to kill your own self, she thought.

So that evening in 1989, Gaynel joined the fight. Margie Richard would remain the leader, and keeping true to her promise to stay organized, she insisted upon votes for vice president, treasurer, and secretary. Gaynel was elected secretary. Cathy Street's Rosemary Brown, who was married to a cousin of Margie's and knew the town's history well as a longtime activist, became vice president. Roberta Johnson, who ran a nursery from her Cathy Street home not far from the plant, was elected treasurer, rounding out the board as the community officially formed the Norco Relocation Committee.

Their dues: $2 for the elderly, $5 for those who could afford it.

Their target: a company so rich with crude it would one day turn $1.5 million in worldwide profit *each hour*.

On the white side of town, many homeowners work for Shell, are Shell retirees, or earn a living by the service industry that caters to the plants along the Chemical Corridor. It's a safe community, with fine schools, proud churches, and long-term neighbors, and most residents there couldn't fathom what Margie Richard and her neighbors in Diamond were getting worked up about. Sure, the plant had accidents, but what plant didn't? Shell was *the* company in the company town, and the other side of Norco stood proud of it, so much so that it could muster the political muscle to elect a Shell employee (and neighbor) to serve on the St. Charles Parish political council. Shell was never shy about highlighting the impact its Norco plants had on the community and country. At the River Road Museum, visitors learn that during World War II, the Shell refinery developed aviation fuel that helped U.S. warplanes defeat the enemy.

At the elementary school rises another symbol of Shell's standing: "Welcome to Norco Elementary. Together Partners in Education with Shell Chemical LP." At the Shell chemical plant, the employee parking lot overflows with USA-made

Ford and Chevy trucks, some with bumper stickers affixed reading: "Real Men Love Jesus." White Norco's front yards are filled with Virgin Mary statues.

The Diamond community's view of the company remained mistrustful, and Shell still offered few jobs to people along Washington, Cathy, Diamond, and East streets. Diamond couldn't afford elaborate scientific studies to prove Shell was making it ill, but practically one of every three children suffered from asthma or bronchitis.

"If we're going to do this we have to do this right and we have to use facts. We have to stick together. We have to stay focused. We have to overcome a whole lot of obstacles, and one is fear, and people who are afraid to speak up," Margie told her neighbors.

Practically every week the Norco Relocation Committee gathered at someone's house in Diamond, and each meeting seemed to grow by one person. Some nights they met at Janice's on Diamond Road, other times at Gaynel's, then at Margie's trailer on Washington or her mother Mabel's house right next door, then back to Janice's or Gaynel's or over at Juanita Johnson's house on Cathy. Margie's mind fixated on business, and she even enforced *Robert's Rules of Order,* the code of conduct local governments adopt, in meetings.

They prayed before each meeting and put aside the small group dues to pay for cardboard and sticks to make their homemade picket signs of protest.

From the beginning, Margie had her neighbors record their list of goals on paper.

Number one on the list was relocation from Diamond. Also on the list:

→ a right to a long life, longevity

→ cleaner air

→ less noise

→ reducing the flare from the chemical plant's smokestacks

→ getting Shell to talk to them like neighbors

Even though Margie was raised in a southern Louisiana where "Colored Only" signs were common and where one-time KKK leader David Duke had led street marches, she urged her neighbors to avoid stereotyping everyone at Shell. "People of industry have children too," she said, insisting that even corporate giants would want to know if their plant was spoiling the air. "If we don't tell them, how will they know?" Margie preached. "There are two sides to everything. People are not talking to people, they are not respecting people, and I don't like to see people mistreat people."

As she looked around at those intimate house meetings, the words of civil rights activist Mary McLeod Bethune came back to her. "Never start out to lose the race." She knew her goals were attainable, but only by breaking down walls, not creating new ones.

Margie and many others in the community believed they needed to move from the blasts and the smells, and that Shell should pay for them to relocate. For years, Shell had bought pieces of Diamond land a parcel at a time, typically when an owner died and the out-of-town heirs sold to the company. The price Shell paid was dirt cheap—land next to a chemical plant is never costly—and Diamond residents knew the going rate wasn't nearly enough to set them up in new homes. Shell had paid a pittance for Helen Washington's land on Washington Street; but now Diamond residents sought to break this precedent and ask for a fee decent enough to set them free from the plant.

Despite the Relocation Committee's growing membership, voices of dissent were spread throughout Diamond's four streets. Margie and Gaynel encountered this resistance firsthand when they went door-to-door to drum up support. Some feared confronting the billion-dollar entity across the way, thinking it foolish to believe the small community could force Shell's hand; others would have been happier to run the company out of town. Some thought Margie, who had begun reading up on chemical and environmental regulations and had even made appointments with researchers and state officials she had buttonholed into meeting her, was out of line speaking as leader of the community when not everyone was on board.

"The biggest hurdle is fear. Fear of industry. Fear of what they could do to us, what they could do to the community," she told her neighbors. "The community people have this negative fear: If we do this, what are they going to do to us? Will they still support the community? Will they still support the elementary school? The fear popped up again—'Oh, you don't do this.'"

It didn't take long for the resistance to drain both Gaynel and Margie.

"They can't move that big old place. They're just wasting their time," some neighbors said, and then, in a quieter voice, "Look at that Margie and that Gaynel."

Margie had also ventured beyond the thick bushes that

separated the dead-end streets of Diamond from white Norco, and those visits likewise stirred defense of the status quo. During trips in white Norco she would sometimes hear, "When are you going away?"

Margie would return the next day or week, to the same grocery or bank situated squarely in white Norco.

"The other side of town and some of those leaders in the plant stereotyped us. There was people saying we were making noise, *as usual,* 'typical Negroes.' All of that gave me strength because I'm not a 'typical Negro,' I'm a human being. This is when all these forces motivated me from within to stand up. I began to speak out in our churches," Margie said.

Soon, the newly formed grassroots campaign was confronted with an additional battle. Shell Chemical Company was seeking a state hazardous-waste permit that would allow it to operate seven hazardous waste storage tanks—three existing and four new—along with two already existing organic chloride incinerators and three new nitrogen strippers. All of them would operate just a short walk from Norco's elementary school.

At public hearings over the viability of the storage tanks in August 1989, a Shell representative detailed the benefits of the company's expansion plans to government regulators and Norco residents. "Few people come into contact with the final products made at the Shell Chemical facility at Norco, yet few people can fail to contact them indirectly in their daily lives. Many products of the plant are in turn starting points for other manufacturers or are important ingredients in preparing the final products of many kinds. A number of plant products find their way eventually into paints, varnishes, plastics, cosmetics, and electronics, and/or are used in wastewater treatment and in the processing of such things as wood pulp, paper, fabrics, and high-octane gasoline components," the Shell official said. "The Norco Shell Chemical plant is the only Shell facility which manufacturers these products."

He closed, "This is another step in our continuing efforts toward waste minimization and sound waste-management practices, which include handling our waste disposal on-site, whenever possible."

Josephine Bering, of 115 Diamond Road, stood up soon after and told the gathering how her fig trees in the shade of the plant had all died and that her health was failing. "I am sick right now from Shell," she said. "Every year I have to leave away from here because of Shell and that odor, and I have asthma."

For years Josephine had been one of Margie's close friends, a crucial support beam who aided her fight. More than once, Josephine told Margie she was going to pray with her and pray for her, and as Josephine stood up for Diamond, a parade of neighbors echoed her concerns.

Huey Wilson, from 206 Washington Street right across from Shell, spoke next.

"If we put something like this in here, we're just giving them open license to bring in all kinds of chemicals," Wilson said. "I hear everybody saying, 'Oh, well, the property is not worth nothing.' . . . We got young kids being born in this community. They can't speak for themselves and I think we ought to speak up for them. We as citizens of the parish and the area cannot sit apathetically around and allow Shell Oil and Chemical complexes to destroy our lives.

"They're applying to acquire a permit to dispose of waste that is death in itself," he added.

Wilson drew a sharp contrast between the people pushing for the expansion and the small homesteaders greeted daily by smoke and smells. "The very people who propose these actions for dumping these wastes in St. Charles Parish do not live here," he said. "They make these plans in super skyscrapers."

Next, lifelong Norco resident Harold Cambre detailed the cancer cluster that sprang up on his tiny block. A cancer patient himself, Cambre said four neighbors had died from the disease. "I'm the only survivor in the five houses in a row," he said. "I can't understand how we could let Shell Chemical build something like this incinerator across the street from a school which several hundred children attend daily."

Five months after the residents spoke, the government had the final word. Six days after New Year's 1990, the Louisiana Department of Environmental Quality (DEQ) and the U.S. Environmental Protection Agency (EPA) approved Shell Chemical's permit request.

Shell wasn't embraced so lovingly in circles outside Louisiana. Four months after the permit was approved, an environmental group named Citizens Fund issued a report identifying ten plants nationwide that the organization asserted had emitted the most toxic pollution in 1988. Four of the ten stood squarely in the Bayou State, and Shell Oil's Norco Manufacturing Complex ranked second. Citizens Fund compiled a separate list of the ten corporate entities it said were the nation's largest toxic polluters; Shell Chemical Company landed at number four.

Shell had always put little, if any, stock in the words of environmental groups, and it dismissed this latest report as misguided. Its permit request had been approved, after all, and most of the people in Norco believed the company was an asset.

In Diamond, where the Norco Relocation Committee was trying to gain momentum, Margie Richard found a key pillar in a longtime friend, a white chemist named Wilma Subra. She had begun working in the Norco area in the 1970s as part of her environmental outreach, and she had got to know Margie then. Subra volunteered her time in Norco, as she does in other hard-pressed communities. Her company, occasionally backed by grant money, stays afloat through small business clients.

Wilma Subra carries herself with the cordial disposition of a school principal, or of the grandmother that she is, and her gentle voice is pure native southern Louisianan. Look past the soft exterior, however, and you'll find a woman who digs for environmental truth with the skill and tenacity of a detective. She's based in New Iberia, a few hours to the west, where her home base was long a trailer with the logo Subra Company, from where she developed a reputation as an expert environmental health scientist. Subra does much of her work from her modest sedan, driving all hours to gather air and water samples.

In Norco and elsewhere in southern Louisiana, her tool of choice is a canister that is something like an oxygen tank. Pulling into an environmental hot spot, Subra opens a valve, and the canister sucks in air. When the job is done, she sends the device off to a lab for testing. She certainly had Shell's attention:

the company paid her the respect due a future MacArthur Foundation fellow whose findings are unflinchingly accurate, even if not always pleasing.

Trekking to Diamond to capture air and water samples, she often found emissions that exceeded permit levels. Each time, she first brought the results to the residents. Then she presented her findings to a group of Shell leaders that typically included Don Baker, the company's community relations manager and liaison with Diamond residents. Baker sported slicked blond hair and crisply pressed clothes, fitting for this tall man with a conservative air.

"They would sit there and listen, and they would say, 'We'll consider it,'" Subra said. "But there was never a response back, 'Well, we agree to do *this*. We agree to reduce our accidents, we agree to reduce our emissions.' And none of that ever happened."

Now, as the Diamond community rallied for change, Wilma encouraged Margie and her neighbors to keep logs of the emissions from the plant, so she could compare them with reports kept by the EPA, and examine whether the government was monitoring the plant's discharges vigilantly. She agreed to help the community build its case for relocation because she believed fiercely that the Diamond homeowners should not live right next to Big Oil and Big Chemical.

"Margie, you're going to die. If you don't get out, something's going to happen and you're going to die," Subra said. "It's really not good for you to live this close to this facility. You really need to think about getting out."

"Yes, I understand what you are saying, but this is home," Margie would sometimes say. "My mom is here."

The 1988 explosion drove Subra's point home even more clearly, and Margie decided to do whatever she could to move all of her family and friends away from the chemical plant.

"After the explosion, everybody should have moved out," Subra said, meaning Norco residents black and white. "People living all along the river are much, much too close. And from an emergency-planning standpoint, much too close."

Subra visited Diamond often, and touring the small community one afternoon, she was struck by how railroad tracks boxed in an area that already offered precious few exits. The Shell Chemical plant stretches for acres on the east bank of the Mississippi River; it's bound on the west by the Bonnet Carre Spillway, on the south by River Road and the Mississippi, on the north by the Illinois Central Railroad tracks, and on the east by the Diamond community. "If something goes wrong, you're stuck," Wilma told Margie and her neighbors, and then she listed other hazards facing their community: "Respiratory problems, difficulty breathing, skin rashes, the kids with asthma on nebulizers."

Her counsel and findings added heft to what Margie already knew. She and the growing Norco Relocation Committee continued to withstand the naysayers and retreated to their trailers and wood-frame homes, where they cut out poster board and began making placards to hoist on their lawns, letting Shell know they were ready for change:

POLLUTION KILLS

By any standard, they faced a brutally uneven fight.

Five

GRASS ROOTS
AND LAWSUITS

For Margie the battle soon became more personal. Ericker, Margie's youngest, had become a sixth-grade teacher, following in her mother's footsteps. Caprice, her oldest, had graduated from Xavier University in New Orleans, the same college Granddad had attended, and earned a degree in microbiology with a minor in chemistry. When Caprice came home to Norco in the early 1990s, she landed a job as a lab technician inside the Shell Chemical plant, becoming the rare child of Diamond on the company payroll. For Caprice the choice was obvious, and practical. She wanted to do graduate studies at Tulane's School of Public Health but could not afford it. Shell's job-training program offered to pay for her schooling while she worked at the plant.

Suddenly Margie feared her activism would hurt her daughter, that Caprice would suffer the consequences for her rabble-rousing mother. It was the lowest point she could recall since the day her sister, Naomi, passed eight years earlier. Margie would, if necessary, walk away from it all.

She turned to Caprice. "How do you feel?"

Caprice was much like her late aunt Naomi and her grand-dad Theodore, bestowed with a quiet confidence and not one for showing off. "Mother," Caprice replied, "we're two different people. I know you. I'm me, you're you. I do my work, and I do my work to the best of my ability. You do what you've got to do."

Her message was spare but clear. Margie would go on challenging the plant that employed her daughter, because Caprice knew how her mother would handle herself.

From that moment on, Margie and Caprice didn't talk of their conflicting missions. Some days Caprice could look out her window at work and see her mother leading a protest for change outside the chemical plant, and Caprice's own boys— Margie's grandsons—were right beside her. Caprice caught the sight, knew her mother was doing this for a reason, and went back to work. If Mom wants to challenge the plant that employs me, Caprice felt, that is fine. The people *should* know if there is a link between the plant and their health, she thought. We won't ever know unless someone asks about it, Caprice believed, an echo of her mother's long-standing credo.

Some in town wondered if Margie was getting lessons about chemicals from her daughter, the Shell Chemical lab tech. Margie didn't dare ask her child for help, believing the only way her activism would take root was if she did the groundwork with her own hands.

On Sundays, as always, everyone gathered at Mabel and Theodore's house on Washington Street for a day of feasting and football. As generations of Eugenes filled the cozy one-story home, chatter over family, school, church, and the beloved Saints carried the day. Not once did Margie and Caprice's conversation turn to the plant.

During the week, away from family and before or after school, Margie continued reading up on chemicals and pollution, attending environmental conferences, and meeting with researchers at nearby Xavier University's Deep South Center for Environmental Justice. She invited regulators from the Department of Environmental Quality to attend some of Diamond's grassroots meetings. Soon, Margie was speaking not only with conviction about the plant, but with an authority that startled the regulators who were paid to watch over it.

Diamond's air still reeked, and its residents, particularly its children and elderly, still suffered from an unusually high rate of serious illness. Shell continued to maintain that there was no link between its plant and the community's health, boasting how its own workforce was happy and healthy. Norco remained a town divided.

For years some of the residents had asked Shell to buy out their homes at a fair price so they could move from the plant. And for years Shell had made offers that residents believed were pathetically low. The company's math always referred to "fair market value." Shell was ignoring one key element of the equation: the plant's proximity had so reduced each home's market value that residents needed more than what their property was worth to live anywhere else. Every subsequent explosion, leak, or scare made the residents' homes worth less.

In Diamond, "fair market value" was a ticket to the poorhouse. Still, a few sold, though not for much.

The Louisiana Bucket Brigade, an aggressive nonprofit group that would later join Diamond residents in their battle with Shell, crunched the numbers some years later. Since the late 1970s, Shell had bought four dozen of Diamond's 269 lots. The average price per lot: $24,350. Brick homes went for the most, about $40,000 apiece. Wood homes went for half that, $20,000, and mobile homes and empty parcels for half of that, $10,000. The Bucket Brigade concluded that Diamond residents would have to go into serious debt in order to move.

When Margie and Gaynel Johnson went door-to-door, they told their neighbors they hoped to convince Shell to buy out their homes at a reasonable price, so the residents would have enough money to move to healthier neighborhoods. "We want a fair market price for everybody," Gaynel told her neighbors, and she said the goal was to get enough money in the sale to move into a brand-new house elsewhere, a "turnkey" equation that made plain sense.

They visited the community's elderly residents, who were particularly susceptible to plant- and refinery-caused respiratory illnesses, early in the morning, at night, and on weekends, beginning each meeting with small talk about family and the weather. Many residents were still angry over the payments they had received after the 1988 explosion, thinking they had deserved more for the damage done to their homes, but few saw how their anger with Shell could do anything to force the company's hand now. If anything, it showed that it was the company, not the people, calling the shots in Norco.

When Margie and Gaynel and any other residents willing

to join them began putting protest signs up in their yards, Shell ignored them. Occasionally a neighbor still had sharp words, afraid of stirring the plant that towered over them.

Margie and Gaynel worked past the anger and urged their neighbors to focus on the simple goal of getting Shell to bankroll their move out of Diamond.

Still, they continued to encounter resistance, particularly from the two Diamond streets farthest from the chemical plant, Diamond and East. Some spread the word that *everyone* would have to sell their homes, that the people had no real say in the matter now that Margie and her friends were taking control.

"No. You don't have to do this if you don't want to," Gaynel clarified. "But if you decide to sell, you will be granted enough money to purchase another house."

Eventually, some of the reluctance in Diamond began to fade, once residents knew precisely what the Relocation Committee had in mind.

Allen Myles, the black lawyer drawn to the Masonic hall not long after the big explosion of 1988, observed Margie and Gaynel Johnson's strident activism, and with a white law colleague named Patrick Pendley, he went about trying to get Shell to settle with the Diamond community. Myles drafted Pendley to the battle in good measure because the men had enjoyed success elsewhere in Louisiana. Both worked from law offices in Plaquemine, about an hour's drive from Diamond, and now they planned to become intimately acquainted with the people and the troubles of their newest fenceline fight.

Pendley, a lifelong Bayou resident, was born in southern Louisiana and graduated from Louisiana State University in the state capital of Baton Rouge. He has a commanding presence but a personable manner, and you can see that he would be convincing, comforting even, with a jury. He's big, not oversized. His law office is too—a two-story wood-frame edifice that looks more like a grand estate than a law office, with handsome framed photos on the wall. It's elegant, not showy, and there his legal staff refers to him cordially as "Mr. Pat."

"You won't find any chemical plants close to a country club in Louisiana," he says with a knowing laugh.

After Myles drew Pendley to Diamond's cause, they visited the neighborhood, and as they spoke with Margie in her trailer, and then walked through the community, they brimmed with optimism. Who could blame them?

Just before they met Margie, Pendley, Myles, and other lawyers had helped settle two cases with striking similarities to what the Diamond community was experiencing.

The first involved the Sunrise Subdivision in Port Allen, about sixty-five miles from Norco. Like Diamond, it is populated by families housed in wood-frame shotgun shacks, most of the homes standing on piers. The three-street, nearly all-black community, built in the 1800s, stood just across from the Placid Refining Company, once owned by a group of brothers from Dallas. The refinery churned crude oil into gasoline and also made jet fuel, but its noxious chemicals unsettled the seventy families who lived so close that they practically stood in the plant's shade. On Christmas Eve day in 1989, a big explosion at the Exxon plant in Baton Rouge killed two workers, so Sunrise residents carried real worry that their time was coming. They griped about the noise, the air, and the lights that seemed to peer into their tiny homes. "You could stand in any yard and spit and hit [Placid's] fence," Pendley told his colleagues back at the office.

Pendley and Myles's simple goal was the same as Margie's was in Norco. They wanted the refinery to buy the people out from under the Placid plant. To help make their case, they hired an engineer from New Orleans to come out and take lead samples from ditches just off the plant. Scanning the lab results, Pendley looked over to Myles. "The numbers are just astronomical," he said.

The company initially fought them, but not for long, not after the readings confirmed what everyone in Sunrise knew. Maybe Placid Refining bosses worried that a lawsuit could close the plant, or maybe they feared liability if a blast did come, but nonetheless the company settled in 1991. It bought the houses of the claimants, paying enough money for dwellers to find new homesteads.

Soon Margie and her Relocation Committee took field trips to Sunrise to visit the land and the plant, and to share picnics and stories with the families that had lived there; and the Diamond community invited the people of Sunrise to join their regular house meetings.

Another case that concluded with a settlement to move a community away from smokestacks also gave Pendley a shot of optimism. This was in Donaldsonville, Louisiana, at the CF Industries plant, which makes fertilizer with ammonia. "You get next to the CF plant when the wind is right, it will bring tears to your eyes," Pendley liked to say. Approximately eighteen families lived in the subdivision next to the facility. Mostly white and poor, a mix of elderly and working-age people, the residents had been asking the plant to move them for years, but got stonewalling and silence. Pendley and other lawyers filed suit, and the settlement quickly followed. Many of the houses were brick veneer on slab, and some of the owners simply had their homes hoisted off the land and moved to a plot of property a good distance away from the smell.

Now twice coauthor to victory, Pendley couldn't help but look at Diamond and see another winner. "It's a mirror image. The same exact situation," he told Myles during the hour-long drive back from Margie's trailer home. With one exception—Shell had long had in place its "fair market value" buyout program, which targeted the two streets nearest the chemical plant in Diamond, Washington and Cathy, and a third street, Good Hope, in the white section that stood nearest to the refinery. The company consistently let the community know it would be open to buying properties on those streets if owners were inclined to sell. The other companies Pendley had challenged had no such buyout program in place until the lawyers came knocking.

"If you had laid out the three scenarios and asked which would settle, I'd have hit Shell like a duck on a June bug," Pendley observed. "We're not talking about $10 million. I doubt we're even talking about $5 million."

Yet Shell's buyout wasn't as promising as it seemed on the surface. "You couldn't buy a decent mobile home for that," Pendley noted. "For people who were living there, they just couldn't do it."

Pendley and Myles went into Norco and talked with real estate agents to get a fuller picture. What would it cost to buy a 1,500-square-foot house, wood frame on piers, with a car-port—in Norco but away from the plant? They took the numbers to Shell, thinking it was pocket change.

Shell did not budge. "We'll buy every one of them," a Shell lawyer told Pendley. "For fair market value." Not one penny more.

The company believed its stance was fair. Why overpay for modest houses abutting the chemical plant? Shell certainly wasn't about to set a precedent of forking over more money than its appraisals suggested. "How could you justify giving some folks more money so they could move elsewhere? Is that good business sense?" one Shell manager explained.

Driving back to Plaquemine, Pendley and Myles rolled the company's response like a Rubik's Cube, but they could not solve it. Why not spend a fair price, move people from the fenceline, and avoid more expensive future liabilities?

Furthermore, Pendley reasoned, current Louisiana regulations legally forbade homes to be this close to industry, mandating a two-thousand-foot buffer zone. Yet that law was adopted years after the Shell Chemical plant and Diamond homesteads rose side by side, with properties along Washington Street a mere twenty-five feet from the fenceline. While the two-thousand-foot law sounded good on the books, it did not apply to those neighborhoods constructed near industry *before* the reform was approved. The Diamond community had arisen at a time Louisiana lawmakers were giving little, if any, thought to the people being hugged by Big Oil and Big Chemical.

One afternoon in mid-1993, Pendley picked up the phone and called his opposing counsel at Shell: "We're going to file suit."

On October 5, 1993, *Richards v. Shell Oil Company et al* was filed in the courthouse at the Parish of St. Charles, on Highway 18 in Hahnville, just down the road from the chemical plants that line lower Louisiana. Someone initially labeled the case *"Richards v. Shell,"* and that typo clung to the legal dispute through the four long years of courthouse maneuvering that would follow, but there was no doubt that Margie Richard was lead plaintiff representing the interests of like-minded residents.

"Nature of Injuries?" the legal form asked. "Claim for negligence, reckless and wanton misconduct, strict liability, nuisance, trespass, battery, intentional infliction of emotional distress, negligence in the conduct of ultra-hazardous activity." That about covered what Diamond lawyers believed should deliver a verdict awarding enough money for residents to reach their ultimate goal of moving out.

The twenty-five-page complaint said that the "air breathed by the plaintiff . . . is contaminated" and that Shell "will continue to emit substantial quantities of hazardous waste and chemicals onto the plaintiff's land."

Harm to the community? "Physical, economic and psychological," the suit maintained, and it detailed how the plant's emissions dropped the market value in Diamond and tainted the community's air, food, and water. "Indeed, there is no safe threshold level of exposure to many of the chemicals emitted from the Shell facility."

The lawyers closed with a section sure to catch Shell's eye. They wanted the courts to shut the chemical plant down, "enjoining Shell from further use of said facility . . . and decreeing that in default of Shell removing said facility, that the Sheriff of St. Charles Parish be directed to remove the same at the expense of the defendant, Shell."

With those legal words, the community's battle with Shell was joined.

Shell's corporate and legal strategy was simple. "If they felt they were not wrong on these issues, they would fight the lawsuit," one Shell official said.

Fight it indeed.

Shell would come back, hard, and the Diamond community and its attorneys would quickly learn firsthand what it's like to grapple with a giant. Soon enough, it was their legal maneuverings that would come into question.

Six

"A FAST
ROUTINE"

Two weeks and a day after Shell was served the legal paperwork of *Richards v. Shell*, the oil company was back in the courthouse, though this time in the federal court of New Orleans, in the matter of the explosion of 1988.

Five years after the Norco sky was enveloped by a blanket of yellow and red, federal judge Henry Mentz Jr. scanned the staggering legal numbers triggered by the blast's cost in pain, suffering, and destruction. More than seventeen thousand plaintiffs had been brought together in one class-action case, described in the New Orleans *Times-Picayune* as "one of the largest ever involving a single event."

Mentz signed his approval: $172 million for precisely 17,146 claimants. Since the hundred-plus attorneys involved received a sizable cut—$30 million in fees and $13 million in expenses—$129 million was left for the plaintiffs. Individual payouts typically ranged in the thousands, depending on the severity of loss, and people closest to the refinery got an extra bounty. Those within one mile pocketed at least $12,000, within two miles $9,000, and within seven miles at least $7,500. Many got much more, but the average payout amounted to just over $7,500 per plaintiff, which came on top of fees Shell and insurance companies had paid in the wake of the blast to cover immediately provable damage.

Early that December the people began lining up outside a LaPlace office building, where the sign taped to the window told them they'd come to the right place: "Shell Norco Explosion Settlement Identification Center." One by one, the plaintiffs stepped through the door, sat before a video camera, pulled out their ID cards, then signed this final batch of legal paperwork, the cameras recording every step to make sure phony plaintiffs didn't suddenly start appearing. Each visit took just ten minutes or so, and the checks were cut shortly after, three days before Christmas, a rapid conclusion to a case that could have clogged the courts for decades. Banks stocked their coffers to ensure they had enough money on hand to cash the thousands of newly arriving settlement checks, and local stores glistened with holiday goods for sale, but for some plaintiffs, the settlement check simply went to help make ends meet.

These final class-action payments came after Shell had already settled with the families of the seven deceased workers trapped in the plant that early morning.

Yet behind the mammoth numbers of the Christmas Eve settlement was another story altogether, one that gained currency anew as the checks were being cut that December day in 1993, and one that raised the distinct specter that Shell had taken advantage of people in shock in the aftermath of the 1988 blast to minimize its ultimate payout.

In November 1988, six months after the explosion, most residents were just getting their initiation into the legal world's snail pace of justice. As frustration mounted over rumors that some plaintiffs would be dead by the time the final settlement check arrived, initially estimated by experts to come a dizzying seventeen years into the future, word went out that Shell was making an effort to help those who needed money urgently. The company set up an office where residents could pop in, sign a form, and walk away with $1,000. The office buzzed with activity and Shell lawyers who made the process easy to navigate. The pitch was simple. If you believed the company and insurance adjusters had been at work fixing your house properly and were treating you fairly, just sign the form and grab the cash. More than eleven hundred people did.

Among them was Margie Richard, who felt the claim she submitted for repairs on her mobile home had been going relatively smoothly, and who confronted the reality that daughter Ericker's college tuition was due at Southern University in New Orleans. Christmas 1988 would be here in a month, and the bills were piling up.

She had heard the news one day while at school—Shell was setting up an office in the white section for anybody, around November, to come down and sign papers. When Margie showed up, she saw a long line and clusters of lawyers.

"It was an opportunity around Christmastime to get the tuition paid," she said. "I didn't think any further. I was in my mobile home by then. I figured they were fair."

Margie signed the papers and took the $1,000.

In doing so, she and dozens of her neighbors, after receiving counsel from Shell representatives, gave away their right to sue by signing a contract that included those typical and definitive legal catchphrases.

Five years later, when the huge settlement was cut, Margie was among those on the outside looking in, no longer able to collect damages over the 1988 blast. A conservative accounting shows that Shell shaved nearly $7 million from its ultimate payout by giving smaller settlements to these eleven hundred residents up front.

"I did what I had to do at the moment at the time. I'm not a lawyer. But I learned something," Margie concluded.

As the final checks were being issued in 1993, Margie and a lawyer journeyed to One Shell Square, the fifty-one-story skyscraper adorned in Italian travertine and bronze glass that towers over New Orleans as the single largest office building in the downtown business district. Modeled after One Shell Plaza in the company's U.S. headquarters in Houston, the gleaming, boxed tower had arisen with great fanfare in 1972, one year before a Shell Chemical explosion quietly took the lives of Helen Washington and Leroy Jones. Sitting in this stout symbol of power and privilege, Margie Richard turned to a member of Shell's legal team.

"It's a shame you are doing those things. It was trickery," she said. "It was a fast routine. Why would they do that at a time when they knew people were shaken up?"

The company didn't believe it had engaged in trickery at all. In a time of chaos and pain it had offered payment to people desperately needing it. Anyone could sign—or not sign—the form and walk away with—or without—$1,000, an offer that had been made with court approval. That some may have missed the fine print was certainly not the company's fault. "That's their feeling and I understand their view. I didn't think Shell took advantage of them," said one longtime Shell official. If anything, the $1,000 had been a generous offer of *additional* money to people who were already pleased with the progress of their repairs.

Yet now, with the federal judge finalizing the big settlement, many Diamond residents thought of how Shell had enticed them with easy money, and wondered whether they had been taken.

Others took a different view. "I think it's a lot of people trying to get something for nothing," one longtime Shell employee concluded concisely.

Seven

RICHARDS
V. SHELL

Margie Richard took a seat in the Shell lawyer's office in Destrehan on March 28, 1994, flanked by Pendley and Myles on one side, and two lawyers and an investigator for Shell on the other. She was ready now to give her sworn statement; the court stenographer sat poised to record her words.

"And your sister is dead now?" the Shell lawyer asked.

"Yes," Margie replied.

"What did she die from?"

"Sarcoidosis."

"I'm sorry?"

"Sarcoidosis," she repeated.

"What is that?"

"A lung disease."

"What is it that you see in the air?" the Shell lawyer wanted to know a few minutes later.

"Sometimes this white cloudy stuff, it smells like bleach. Sometimes some little black smut-looking stuff that will show up on any white surface, on my trailer, on the steps or the door," Margie explained. "If your air conditioner is on, you get this choking cough, and you say, 'Oh, I smell something.' You get up, you go outside, and it like strangles you to death."

"Can you tell me what evidence you may or may not have with regard to permanent injury to your land, to your father's land?"

Margie answered that she could show the lawyer her dead flowers and dying garden. "On the left side of where I live, everything dies," she said.

"Now in your petition you've talked about reckless and wanton misconduct," the Shell lawyer said. "What evidence do you have that these guys actually did that?"

"We wanted to see a representative from Shell . . . before all of this was filed to just come talk to us or do something, and there was never any response," Margie said.

Finally, she had got calls—at school as she was teaching class. "It wasn't the right time or the right place," Margie told the lawyer. "Why now, when we tried to talk to you all before this? Because, if anything, we wish this could have been done without all of this, but nobody got in touch with us, like they didn't care."

Margie testified that some of the calls to her school had come from Shell's Don Baker. When Margie or her neighbors reached out to Shell, the company always directed them to Baker, and Baker alone. Don Baker was the public face of Shell, but some in Diamond said they rarely saw his face in their community.

"But somebody called you when you were at work?"

"After the lawsuit was filed," she answered. "I told them I could not communicate with them then. . . . I said, 'See me and the lawyers.' But he never did, they never did come."

"Where would you go if you were moved out of Norco?"

"All I want is out, but I would move somewhere I know where I would not hear the consistency of the aggravating noises and the smells," Margie explained. "I'd like to see my parents live out their older days without what they're going through now."

"So what you're telling me is you don't really have any location that you would want to move in mind at this time?" the company lawyer pressed.

"I go where some clean air is," the community's lead plaintiff testified, and she spoke of how she had witnessed family and friends suffer from cancer, sarcoidosis, and other ailments. "They say the plant is not at fault, but then, who knows? And that's not oil; that's chemicals. But those who saw that fire dance across that street, it's just every little noise go off now, it's a fear that how long it will be before the pipes right in front of my house go?"

"Do you have any particular evidence that would support any claim that you and the two hundred and eighty or so odd members are being treated any differently than, say, people that live on the north side of Norco?"

"The only evidence that we have that may not be concrete

is what we walk through and live through daily," she replied. "It's like we're forgotten. It's like, let us die. I mean, look, that noise and that flaring is bad. It shakes your house. The evidence is our dishes rattling in the cabinet, doors shake.

"The evidence, it's like people on our street die."

With those words, Margie Richard rose from her seat in the Shell lawyer's office, her 112-page pretrial deposition now concluded, and the real legal battle of *Richards v. Shell* just beginning.

This is the long road toward justice few see in movies or on TV, where the dramatic courtroom scene is so often the heart of the story, the Perry Mason moment when a star witness crumbles under fire, or the Jack Nicholson "You can't handle the truth!" shout-down to Tom Cruise. Lawyers and investigators practiced in the art of civil action know that the early stage, well before the final act, is where lawsuits are often won, and sometimes lost.

Margie was lead plaintiff, but she was not alone in detailing Diamond's ailing health, and her neighbors' concerns also shaped the core of the community's suit.

A woman from Diamond Road cited her "skin discoloration plus rash, sinus, dizziness excessively [*sic*] ringing in ear, blurry vision or eyes, night blindness, chest pains, shortness of breath, arthritis." A neighbor on East Street reported "strange itching skin rashes appear all over body out of nowhere, scalp itching cracking discoloration, flaking, peeling pus. Deep pockets under skin, red painful swelling. Sinus trouble so bad that cannot smell something burning. Ear ache, pain. Hearing loss ringing in the ear sometimes eyes, burning, watery running, irritated double vision and night blindness [*sic*]."

Under oath during more of the case's pretrial questioning,

a Cathy Street resident complained of the loud noises, and how the plant's chemical releases caused her eyes to burn and her breath to fall short.

"Have you ever gotten any medical attention whatsoever because you were nauseated or couldn't breathe?" a Shell lawyer asked her.

"No. I couldn't afford it."

The lawyer followed up. "Why don't y'all move, if it's a problem as far as the operations of the plant?"

"Can't afford to," she replied.

A man who had lived in Diamond for thirty years before finally moving out a few years earlier reported suffering muscle problems, a heart attack, and Parkinson's disease. "That steam and all the vessels and everything was right in my front yard. Everything they clean out was right in the front yard of my house. We didn't know at the time that the steam was blowing right into my kitchen, directly into my kitchen."

"Now, the complaints that you have against the Shell Oil refinery and Shell Chemical, what are those complaints?" the Shell lawyer wanted to know.

"My complaint is the breathing condition," the long-timer explained. "From my nose, head, keep me all congested and everything. I got some complaints about being frightened so much, because I would show up and complain every time. Every time something happened, I was over there."

Doris Pollard, Margie's cousin and the daughter of the late Inez Dewey, had an adult daughter who had sarcoidosis, the same disease that killed Naomi. Doris walks with the aid of a cane, and in 1994 she was hospitalized for pneumonia. "I suffer trying to catch my breath. It seems like in Norco I can't breathe. Whatever coming from the plant cut my breath," she told the Shell expert before trial.

Doris's dreams about death disturbed her, forcing her to go as long as two days without sleep. She could not erase the day in 1973 that she saw two neighbors die, or the memory of how she and her mother were saved from the blaze only by the heroics of a neighbor who saw them trapped by the fence. Her gravest fear now was that the whistle would go off at the plant, alerting the town to another catastrophe, but that she wouldn't wake up, and would burn to death inside her home.

Pendley and Myles documented the stories from a community that until now had had little audience for its oral history, and then the lawyers turned to a slew of hired professionals to translate the pain and suffering into a legal framework. In October 1994, a Boston professor of pathology concluded that the people of Norco "have been, are and will continue to be at significantly increased risk for the health problems . . . because of the unusually close proximity to the site." A clinical psychologist from the Louisiana Polytechnic Institute examined the lawsuit's fourteen core plaintiffs and recorded their nightmares, fears, and flashbacks, the residents' stories now forming the basis of *Richards v. Shell*.

Shell was busy engaging its own set of specialists, and its lengthening roster of high-powered doctors, experts, and consultants filed a flurry of reports in the official court record leaving room for just one conclusion: The Shell chemical plant had no impact whatsoever upon the lives and the health of the people residing on Washington, Cathy, Diamond, and East streets.

In November 1994, eight months after Margie Richard's

plainspoken deposition, a Tulane University Medical Center Ph.D. filed a report to Shell that came to a strikingly different conclusion than she had concerning the Diamond community. "The long list of general complaints is essentially a shopping list with no factual, scientific basis," the Tulane Ph.D. wrote. "It is therefore my professional opinion as a toxicologist, that the operation of the Shell Norco Petrochemical Facility presents no concern to health of neighboring residents. This is consistent with the absence of specific health complaints by the residents."

The same month, Shell received a report from its hired noise expert. "The fenceline noise levels did not appear to present an environmental noise concern and the levels generated by the chemical plant, and reaching the fenceline, were no greater than noise from general community activities."

Shell's pollution expert's report arrived at the same time, and it too asserted that the plant was doing no harm. "The results of the studies I have conducted have led me to conclude that the Norco Manufacturing Complex does not violate any ambient air quality standards. This conclusion is based on detailed study of the plant sources and their emissions," the expert wrote.

A year later, as the case dragged on, another Shell expert weighed in with a new report adding yet more heft to the company's mounting defense, concluding that the chemical plant's environmental-protection program was in fine order. "Environmental personnel have a good grasp and understanding of environmental regulations and good management practices," the Shell witness asserted.

The reports made it clear to the plaintiffs' lawyers that

Shell would not only raise a spirited defense; it was out to show that the community had no case at all. Shell heard the residents' complaints about their ailments and replied that there was no evidence at all to support their claims.

As the legal plotting and dueling expert opinions continued inside the courthouse, Margie Richard, Gaynel Johnson, and their tight cadre of organizers pressed forth outside it, in Diamond.

They called community meetings at the Good Hope Missionary Baptist Church on East Street and handed out old-fashioned flyers headlined "ARE YOU CONCERNED ABOUT YOUR HEALTH?" They continued knocking door-to-door to gain more momentum for their cause.

The idea was to target their neighbors in Diamond, but also any residents along the Chemical Corridor likewise worried about pollution. When neighbors huddled at the parish library in Destrehan, Margie reiterated the importance of the struggle. "We have been nearly blown off the earth," she reminded them.

In white Norco, the view had still not budged. Margie was bringing a case against the very company that helped the town run, the residents thought, and they were angry that this one pocket of Norco was trying to tell the world their town was unsafe.

Sal Digirolamo was among those white Norco residents who continued to wonder what Margie and her neighbors were clamoring about. Shell employed Sal, his father, two sons, five brothers-in-law, and too many friends to count. Sal had worked at the chemical plant for forty years. His job for most of that time was to keep the instruments that measure the plant's flows and temperatures running smoothly. He designed them and then repaired those needing work. Toward the end of his career he worked on inspections, solving problems, and he'd begin each day walking through the chemical plant looking for trouble spots. After the rounds were done, everyone sat down at a big table and set a plan of action. It was a team, and Sal was damn proud to be part of it.

In 1993, just as *Richards v. Shell* was filed in court and as the massive refinery-explosion lawsuit was being settled, Sal told his wife, Deanie, "There are going to be some big changes. You have had the house to yourself for forty years. I'm going to be part of that. The day is going to be changed drastically for both of us. You want to watch TV in the day, I'll be there with you now," he said. At Sal's restaurant, his favorite eating establishment (which coincidentally shared his name), he sat down with three dozen colleagues, family, and friends and ordered his usual, eggplant Parmesan, to honor his time at the plant and celebrate what would come next in retirement.

With more time on his hands after forty years with Shell Chemical, Sal, president of the Norco Civic Association, decided to investigate for himself what the uproar was all about over in Diamond.

Sal always felt he and Margie could talk out their differences. "We argued like heck on both sides of the fence. I still give her a good handshake when I see her. She knew how I stood," Sal explained. "I felt sad they were doing that to a community that was close-knit, that was healthy. It kind of hurts when somebody is talking about a community that we love."

Sal, like Margie, liked to see things with his own eyes, and like Margie, he was not afraid to cross the barrier between black and white communities. Sal would even sit in on some of Diamond's grassroots meetings. He didn't feel shunned, and one day he and Margie sat down over a Sunday meal at the Subway sandwich shop in Destrehan.

"I don't agree with you, Margie. The reputation of our community is at stake, and you ought to look at it real hard," Sal told her. "Diamond is part of Norco."

Margie took her turn, telling her crosstown neighbor of the sickness that had taken her sister's life, had nearly killed her daughter, and had made so many of her neighbors ill.

"Margie, I don't observe that," Sal replied. "I'm not doubting your word. I'm just saying if it's coming from the chemical plant more of us should have it."

Margie wasn't just fighting Shell; she was fighting a way of life in a small town that liked to keep to itself. It took one piece of gumption to challenge Shell, and then another to press on as the town surrounding the plant looked down on her too.

The stress led Margie to turn more frequently to her faith for strength and guidance. In the predawn dark she opened her Bible and ran her fingers over the verses, as if she could hold them. "For I will defend the city to save it," she read, highlighting each word with a bright yellow marker. "God gave me for Norco," she wrote back in the Bible's worn margins.

With the legal maneuvering in full bloom and the grassroots campaign blooming with it, another battle surfaced for Margie. This time it came from her employer, Harry Hurst Middle School, which had begun receiving calls for Margie from Shell after the lawsuit hit the courts. Margie took the calls, told Shell's Don Baker she couldn't talk now, and wondered why the company hadn't called before, or why it hadn't simply walked to her trailer some twenty-five feet from the fenceline and sat down with her when she was not at work. She wondered most of all why Shell was calling her at a public school in the system the company was so proud of supporting.

Once-close colleagues suddenly began turning away when Margie passed in the halls, and one day her school principal pulled Ms. Richard into his office and told her it appeared that her interests lay elsewhere.

"What I do in my spare time is my business," she retorted.

One morning, just a few weeks before Margie's lengthy pretrial deposition, the school principal himself made an unannounced visit to her Louisiana history class for seventh-graders, the one that began at 7:50 each morning. The principal observed the lesson that day, and his written report noted that "Ms. Richard has established a good rapport with most of the students," and that her middle-school class that morning included not a single disruption. Some kids were off task, and the lesson plan veered from what Ms. Richard had initially planned, but the class sounded nearly ideal considering it was a group of twelve- and thirteen-year-olds. When students asked a question about something they should already have learned, the teacher told them, "You must think ahead!" and the students replied positively.

The principal's written report also included a detailed dissection of how Ms. Richard had apparently inadvertently given a wrong date at one point, and how she mispronounced one word, "famine."

Based on the flaws documented in his report, the principal suggested an "assistance plan" for the teacher. He neglected to take himself to task for his own misspellings in the one-page write-up of Ms. Richard: "The students then completed a map assignment which they had begun previosly [*sic*]," he wrote; he used the misspelling "ocasional" a few sentences later; and in the section involving the teacher's instruction that day, he wrote: "There were three inaccuacies [*sic*] presented in today's lesson."

Soon the principal called Margie back to his office and gave her an assignment to write a report about a subject she believed had nothing to do with her class teachings. Margie was to listen to a series of five or six educational tapes on the techniques of teaching and to write reports directly to the principal about them. This extracurricular assignment was to be completed with the Thanksgiving holiday just a month off. The principal told Margie she just didn't seem to be as effective in class as she had been before, and she needed this brushup to regain her standing.

"No, I'm not going to do it," Margie told the principal. "To me that was punishment; it was Thanksgiving holiday," she later explained.

"You're fired," the principal told her.

"Sir, you didn't hire me and you can't fire me. My record speaks for itself," Margie said. "The reason you're doing what you're doing is because of what I'm doing against the company."

The principal did not reply.

On October 21, 1994, while her court case against Shell was in full swing, the principal formally suggested that the twenty-nine-year teaching veteran be fired. "I am recommending termination of employment for the above named teacher due to failure to successfully complete her third stage assistance plan," the principal wrote. "The employee is a tenured teacher who did not fully participate or complete the agreed upon assistance plans. It is regretful that such a recommendation is necessary at this time but [it] is due to failure on her part to successfully complete all three assistance plans."

A week later Margie wrote to the St. Charles Parish Public Schools, explaining that she had major health problems and that the medication she was taking made her sleepy. She also cited the "stress and heart problems" she had encountered in her family life.

Her son-in-law Allen Torregano, husband to Caprice and father of Margie's oldest grandchild, Chris, had recently died of complications from a failed kidney. The children in Margie's classes knew that this loss had deeply affected her. "I'm sorry about what happened. It happened to me too so I understand how you feel," one child wrote in a handmade card decorated with a flower. "Dear Ms. Richard: Sorry to hear about your son-in-law. But we need you here at school. I miss the way you teach. But, I also miss you," wrote another student.

"When I was placed on observation it was a crucial time after the death of my son-in-law," Margie wrote to the school system that was now cracking down on her. "However, knowing the importance of teaching and my responsibility to the students I placed them first. When I attempted to complete the assigned work given to me by [the principal] I felt punished. Additional stress and health problems developed."

She steadfastly refused to write the report on the teaching tapes, telling a committee that was brought in to assess her situation, "I'm sorry, I'm not going to do it, because I feel it was a conspiracy."

The pressure kept building, and Margie became miserable,

suddenly finding herself in a scrap with an institution she had long adored, but she stuck to her convictions. The St. Charles Parish superintendent put the firing on hold the next month, November 1994. "Before Ms. Richard begins her sabbatical leave for the spring semester 1995, discussion will be made at that time regarding her future in this school system."

Margie continued to fight the termination, and it was officially overturned, but this unexpected tussle had sapped some of her energy at a time she needed no distractions. She felt the school was giving her an ultimatum: Do the tapes or lose your job. Margie decided to go out on her own terms. Before she did, she wrote a note saying she was leaving with no hard feelings toward the principal.

In June 1995, the St. Charles Parish School Board accepted Margie's request to retire, thirty years after she had become a schoolteacher and after throngs of young children had passed through her classes. "On behalf of the St. Charles Parish School Board, let me wish you happiness in your retirement and thank you for the contributions you have made," school bureaucrats wrote to her.

Not long after her official departure from the St. Charles Parish schools, she enrolled in night Bible college classes. With the drawn-out legal case finally heading to the courtroom, Margie Richard would become a theology major.

As she officially exited the public schools, Margie thought back to the question Lamar, the serial skipper, had posed to her in class years earlier. "You ever thought about going around the world speaking to people?"

You know what? Margie Richard thought. That Lamar is onto something.

That year she traveled to San Antonio, Texas, for a regional environmental justice conference hosted by the Environmental Protection Agency. When the public was invited to weigh in on any environmental concerns, Margie rose and held the crowd captive with a story about the explosions, deaths, and sickness in Norco, Louisiana.

Among those listening intently was Samuel Coleman, director of the EPA's Compliance Assurance and Enforcement Division for the southern United States, based in Dallas. "She is just a very compelling orator, and understands how to keep the crowd spellbound," Coleman said. "And she can tell the story in a way no one else can."

Coleman was so moved by her words, he decided to see for himself what was happening. When he visited the four-street Diamond community not long after with his staff, he recognized instantly the potential peril of having homes so close to churning industry. "There were obviously some issues, from my initial visit down there," Coleman said. He left town with a vow to return, and a plan to steer the slow-moving vehicle of bureaucracy toward Diamond, Louisiana.

Eight

PROOF

Although Diamond residents had believed for decades that the plant was filling their air with toxins that harmed the health of their community, they never had any medical studies to support this belief. But with the community's civil lawsuit soon coming to trial, that evidence was finally in hand.

The case of *Richards v. Shell,* initially brought in state court in 1993, was shifted over to federal court at Shell's request that same year. But on January 30, 1996, a federal judge shipped it back to state court, in the Parish of St. Charles, at the plaintiffs' request, and there it would stay. It soon was no longer a class-action case, but a lawsuit led by fourteen "bellwether" plaintiffs, spearheaded still by Margie Richard and including a core of her close neighbors.

Back in state court, the judge assigned to oversee the case

happened to be the father of one of Shell's lawyers, and he was quickly removed, at the company's request. The judge who took his place had kin who had been schooled in Margie Richard's classes. In the legal world of the Parish of St. Charles, it was hard not to trip over someone you knew in school or church or maybe even in your own home.

In April 1997, as the plaintiffs' lawyers were busy filing their fifth amended complaint on the eve of trial, a piece of potent ammunition arrived. Researchers from the Deep South Center for Environmental Justice, based at nearby Xavier University in New Orleans, had prepared a questionnaire headlined "Norco/Old Diamond Plantation Health Survey." The survey included a long list of ailments, and Diamond residents were to check off the specific illnesses that plagued their homes—asthma, bronchitis, pneumonia, lung infections, chest pains, congestion, and the like. They also were to log the number of emergency-room visits family members had made, and detail how many inhabitants were forced to use a breathing machine or inhalator. Gaynel Johnson and other Diamond activists went door-to-door helping hand out the surveys, and one by one the families logged the sicknesses in their homes.

The finished Xavier study read like a troubling report card of community health. Based on the results of the questionnaires filled out by forty-seven families, Diamond's grades were failing.

Titled "Community Health Survey: Norco/Old Diamond Plantation," the report was a portrait of long-timers (more than three-fourths of the respondents had lived in Diamond for twenty-five years) with little means—just 9 percent had household incomes over $30,000. Most were classified as "exceedingly poor," with nearly half reporting annual incomes below $5,000.

Residents were given a series of 139 questions that gauged individual health problems. Over 35 percent of husbands and wives said they had chest pains or respiratory congestion, and their children—ranging from toddlers to adults—reported troubling rates of asthma (35 percent) and bronchitis (28 percent). More than one-fourth of the people reported suffering itching, burning eyes. Also common were blurred vision, ear itches, ringing in the ears, sinus infections, and persistent coughs.

The Xavier numbers themselves were daunting, but the stories later recorded by the Deep South Center conveyed the personal toll.

Catina Hill lived a few feet from the Shell fenceline, and her six- and ten-year-old children suffered skin rashes, asthma, and upper respiratory problems. Their childhoods were cluttered with painful anecdotes that children living far from industrial corridors rarely encounter. Janice Darensbourg's teenage son, Larry, suffered allergies the family believed were triggered by the chemical plant's "flare burns," which forced the boy to hustle inside for cover.

The elderly suffered equally. Mary Hollins, a Norco resident for thirty-seven years, had trouble breathing. Washington Street's Hazel Davis, a retiree who ran errands for the neighborhood's sick and served as vice president of the Norco Relocation Committee for a time, would blame a Shell lime spill for the eye infection that took her to three doctors for treatment. Wendell Eugene believed the plant pollution was behind his asthma.

The stories came from every corner of Diamond.

"Perhaps most alarming is the number of respondents, including over one-fourth of the wives and children, who reported they had to go to the emergency room because of respiratory problems," the study said. "Also common were pneumonia, the use of an inhalator, and the use of breathing machines by children. . . . The respondents in this survey reported numerous health problems, and they believed that the environment in their community was leading to specific health problems."

The most pressing finding: "There is a significant negative relationship between years living in the community and reported health."

Medical experts drawn to the community's crisis said not only that the rates of illness were troubling, but that the quality of the air the residents inhaled was too.

Chemist Wilma Subra had conducted extensive studies on Norco's air going back decades. In 1997, Subra was back in Diamond, where she found that the chemical plant spewed methyl ethyl ketone (MEK), which irritates throat, skin, nose, and eyes; toluene, responsible for developmental and reproductive problems; and other toxins. Subra knew this was further proof that a community should not be this close to industry.

Her long-standing research had largely paved the way for the work Xavier was now completing. Subra identified the pollutants in the air, and Xavier quantified their effects on health.

Whenever they met, Margie kept Wilma posted on any developments in the legal case, which was still in the courts. "If there is anything I can do to help, let me know. I have all this information," Wilma told her.

Margie agreed to pass whatever Subra had collected to the Diamond community's legal team.

The Xavier study was dated April 1, 1997, four months before the community's lawsuit would be presented to a St. Charles Parish judge and jury, and Subra's report came that same year. Both offered potentially enriching evidence, supplementing the Diamond community's many anecdotes with more systemic proof of the harm in the air.

Civil lawsuits hinge on hundreds of decisions, some small, some not, made by lawyers and their clients on the road to that ultimate dramatic moment in the courtroom, where justice finally turns.

Many of these earlier decisions turn on which evidence to include and which to ignore. As the case of *Richards v. Shell* finally headed to its final destination, the Hahnville courthouse, the community's lawyers made some puzzling decisions.

The jury, soon to be seated, would not hear of the Xavier study, because the plaintiffs did not present it among their case evidence. Nor would the twelve men and women of St. Charles Parish hear from chemist Wilma Subra. The lawyers, believing they had ample evidence already to win the case and emboldened by their previous successes in Louisiana, would not call her or Xavier's researchers to the stand.

"Nobody ever asked me anything," Subra remarked. "Nobody asked for any of the information that I had."

Nine

IN THE
COURTROOM

In August 1997, twenty-four years to the month after Helen Washington and Leroy Jones perished in the streets and with the activists' lawsuit finally coming to trial, the Diamond community's lawyers framed their case.

"Throughout the years, Shell Oil has continually expanded the physical plant of the refinery and has enveloped the community and its residents," Myles and Pendley argued in court papers that month. "As a direct result of the expansion and operations of the Shell Oil refinery, the plaintiffs have been continually exposed to noxious odors, excessive levels of noise and bright lights during the night time."

Four years after the lawsuit was filed, the lawyers presented the specifics of their legal attack: that the Shell Chemical plant, erected more than four decades before on the grounds of a former slave plantation, had become a legal "nuisance" in violation of the Louisiana Civil Code, articles 667, 668, and 669.

"Together the three articles establish the following principles:

"No one may use his property so as to cause damage to another or to interfere substantially with the enjoyment of another's property.

"Landowners must necessarily be exposed to some inconveniences arising from the normal exercise of the right of ownership by a neighbor.

"Excessive inconveniences caused by the emission of industrial smoke, odors, noise, dust vapors and the like need not be tolerated in the absence of a conventional servitude."

They closed by arguing, "Plaintiffs are entitled to be awarded injunctive relief and damages as a result of being exposed to the constant nuisances created by the Shell Oil refinery." The papers mistakenly referred to the refinery, but in fact the suit actually targeted the chemical plant.

Behind the legal language stood a striking story that the jury would soon weigh with great interest. When Diamond residents had first gathered in their intimate grassroots meetings at the dawn of the decade, Margie had them draw up their list of wishes, and the very first one was relocation.

Now that the case was in the courts, before judge and jury, the focus had shifted. The Shell plant was a nuisance, and the community's lawyers were seeking hefty financial damages. Strict relocation was not part of the suit now being heard, but affixing a financial cost to life in Diamond was.

In their previous relocation fights, the lawyers had no need to take this approach. Those cases never went to trial, as the companies avoided the publicity generated by a civil trial by agreeing to pay a settlement that let residents move away. Those companies weren't named Shell Oil, however. With Shell in no spirit to settle and Diamond's case actually coming to trial, the community lawyers decided their best hope for success was to paint the plant as a nuisance, as opposed to seeking a verdict ordering relocation.

Under law, they believed, they had zero chance of getting the courts to order Shell to buy out their clients' properties. So instead of pushing a relocation suit, they sought the next closest thing, a case that could bring enough financial damages to Margie and her neighbors that they, in turn, could afford to move out on their own. Same result, the lawyers hoped, just a different path to get there.

"It was a legal hurdle we couldn't overcome, so all we could do is ask for money and our thought was if the jury would award money to these people for the damage for their property, at that point Shell would say, 'What do you want, what do you need? And let's agree on a plan to move these people away,'" Pendley explained. "We were taking the only vehicle we could take, and that was to get an award against Shell for money."

To sway the jury, Margie's legal team compiled 255 trial exhibits, starting with an aerial photograph of Norco and then quickly shifting to hundreds of incident reports, unauthorized-discharge records, and the occasional emergency response call reports involving Shell's facilities from the late 1980s to the eve of trial.

The litany of "unauthorized discharges" included leaks of allyl chloride, chlorine gas, MEK, sour water hydrogen sulfide, propylene, butadiene, propane, ethane, natural gas, sulfuric acid, diesel fuel, flammable gas, and hydrocarbons, to name a few. Taken together, the chemicals were a potential toxic bomb in the making, capable of burning eyes, sickening lungs, scarring skin, and seriously poisoning the air.

Some of the state reports they cited were as murky as Diamond's air had been on the most brutal of nights. In August 1989, there had been a discharge of "unknown black material—quantity unknown." In March 1993 came an "odor incident—specifics unknown." And in May 1997, just three months before trial, an incident involving "flammable gas mixture—amount unknown."

Pendley, Myles, and other lawyers now working with them had scrutinized piles of papers at Louisiana's Department of Environmental Quality to unearth these reports, and the language they now cited—"amount unknown"—mirrored, in a way, the Diamond homeowners' long-standing claim that Shell Chemical had nearly always left them in the dark.

With the jury now seated, Shell lawyers, in turn, seized upon Pendley and Myles's financial damages tactic. They openly hinted that the Diamond residents had their eyes only on the company's deep pockets, a legal play likely to sit well in a parish where most people were friends, not foes, of Big Oil.

The conglomerate moved other legal chess pieces, arguing, first, that many of the DEQ reports referred to by the plaintiffs involved Shell's oil refinery, not the chemical plant, and that anything coming from the refinery across town was irrelevant.

Shell submitted its own exhibit list, and it totaled 345 items, beginning with item no. 1: a photo of the exterior at 26 Washington Street, home of Theodore and Mabel Eugene, parents of lead plaintiff Margie Eugene Richard.

Margie remembers the Shell people coming out before the trial to snap pictures of the carport, okra garden, citrus tree, kitchen, and dining room, and she believed it was because the company was finally going to buy her parents' parcel, and wanted to see it top to bottom, inside and out, to assess the value. Some of her neighbors believed the same. Now those same snapshots were being introduced as evidence at trial.

"We were tricked," Margie later said. "We thought they were going to buy us out, but they were using it as evidence for the trial. They used it against us."

Shell said there were no tricks at all, that it had come out to take pictures strictly as part of the trial's normal discovery process, not for the purpose of a buyout, and that the community's lawyers were informed and fully aware of what was going on. Matter closed.

Though separated by only narrow Washington Street and a razor-wire fence, the community and company hadn't talked in any meaningful way for decades. So what Shell viewed as routine trial work was eyed across the way with wary suspicion.

Shell's exhibit list also included detailed photos of Margie's trailer next door at 28 Washington—the hibiscus plant, the dining table, kitchen, washing machine, rear shed, even an outdoor table with pecans—and of other Diamond homesteads.

And so Shell went, presenting a visual snapshot of Diamond that included butterflies floating on flowers, hibiscuses in bloom, garden plants, oak trees, magnolia trees, rubber plants, and elephant leaf plants, a portrait the company said dispelled the nuisance claims and proved, quite elegantly, that the residents at the foot of the chemical plant were prospering just fine.

Shell also filed psychiatric reports it commissioned on the core plaintiffs that served to counter the Diamond lawyers' and experts' contention that the people had suffered mental anguish from a life spent hugging the chemical plant.

One Diamond Road plaintiff told the Shell-hired psychiatrist that in late 1995, "an odor came in the school, and a lot of children had to leave with vomiting." The woman "complained of a flare making noise and a blaze in the air. You live in fear all the time. The odors are getting worse." She had been asleep, like nearly everyone else, when the 1988 blast ripped their world, and she had fallen in the commotion, injuring her knee and hip, and the doctors later told her she had arthritis. "I went to Charity Hospital a long time for my nerves," she reported. The woman told the doctor of the pain triggered by the 1973 deadly blast, "when an explosion killed them people. I almost had a heart attack that day." The Shell expert heard her story, and concluded: "She certainly does not have any significant psychiatric problem. Although she attributes some of her physical problems to the alleged environmental hazards created by the petrochemical installation, I am unaware of any medical literature that would substantiate any such chemical exposure as an etiology of cardiovascular disease."

Another plaintiff had reported flashbacks and multiple visits to psychiatrists after the 1988 explosion. "I fell out of the bed, and my back hit the nightstand, and I'm suffering with that today," the resident said. "They said surgery wouldn't do us good." She reported sleeping only four hours a night now, nearly a decade later. "The sounds of the plant make you not want to fall asleep," she said, and the doctor's report went on, "She said she awakens 'choking' because of an odor which 'they lets out every morning at three A.M. This odor has been going on the last three or four years.'"

"Apart from some subjective complaints of being irritated by odors and by the flare, there is nothing in this patient's history or clinical presentation to suggest that she suffers from any kind of psychiatric problem as the result of living in proximity to the Shell/Norco installation," the Shell-hired M.D. concluded.

The photos and psychiatric summaries piled on top of the evidence that had already come in from Shell's other battery of experts, who concluded that the air was not foul, the noise levels were faint, and the people of Diamond suffered little, if any, ill health due to the plant.

And so the trial rolled forth day by day—evidence and counterevidence, objections and sidebar conferences, Diamond residents taking the stand one day, Shell experts the next.

On August 25, 1997, day six of the trial, another member of Diamond's legal team moved to submit certain DEQ letters into evidence, and Shell objected that the number of letters like this should be limited. The judge decreed that the community could submit all the letters it wanted, but only those involving the chemical plant. The emissions from the refinery across town were out of bounds.

At 10 A.M. that Monday, after the legal wrangling had quieted for a spell, the twelve jurors of *Richards v. Shell* left the courtroom and traipsed to Diamond, and there they spent two hours viewing the community and the chemical plant. They looked toward the plant and then back at the wood frames and trailers, and then they carried the mental snapshot back to the trial, and more testimony, exhibits, and legal volleying.

Four days later, Friday, August 29, 1997, the testimony concluded, and with it came dueling theories from the lawyers. Shell's words were blunt. The Diamond residents wanted one thing, jurors were told, and that was to dig into the company's "deep pockets." Margie's lawyers countered that assertion with video footage of the plant's bright flare burns, spewing late at night when they were not supposed to, and providing proof, the lawyers maintained, that the complex was indeed legally a nuisance.

The jurors went home for the long Labor Day weekend, and as they filed out of the St. Charles Parish courthouse, Margie Richard felt unsettled by what she had witnessed the last two weeks.

Her community's legal outcome now hinged on whether the plant was a "nuisance," and whether Diamond plaintiffs should be paid for their suffering with a financial reward that would dwarf any payday any of them had encountered in their working lives. Margie worried that the community's ultimate goal—to move from Diamond—had been buried under the debate over nuisances and dollars. "We have our proof here. But what happened?" Margie asked herself. "Something is wrong with this."

Back at the courthouse the Tuesday after Labor Day, a

Shell lawyer reminded the jurors that the Diamond residents really had nothing to complain about. "We all live in a society that has to tolerate certain inconveniences," he said.

The lawyers now done talking, jurors listened as state district court judge Robert A. Chaisson delivered the final charge to the jury before they were to begin the deliberations to pick a winner to this four-year battle. "In this case," the judge instructed them, "the plaintiffs have the burden of proving their case by a preponderance of evidence.... If you find they failed to prove any essential element of their case by a preponderance of the evidence, then they have failed to prove their case sufficiently to recover." He instructed them that Louisiana law required nine of twelve jurors to agree to one side in order to come to a verdict. "When nine of you are of the same opinion about this case," the judge said, "that ends your deliberations and that opinion should be your verdict."

With that, the bailiff led the jurors back to the jury room.

Two short hours later, they filed back into the courtroom, and when the judge peered at their verdict, his face betrayed no clue as to their decision. The one-sentence question atop their jury verdict form had asked simply: "Do you find that the operation of the Defendant's facility is a nuisance within the legal meaning of those terms as explained in the instructions?"

A large X had been marked in the box for "No" to the question, and the jury foreman read the verdict aloud—ten votes to two in favor of Shell. With that, jurors didn't have to consider the next question, concerning what amount of damages should be awarded to each plaintiff. There were no damages.

The international conglomerate had convincingly beaten back the small community.

"I kind of wish they wouldn't have dropped the relocation issue," one of the jurors told the local *Times-Picayune* after the verdict had been rendered. "Since they dropped that, it looked like they were just looking for the money."

Other jurors worried that the Diamond residents could be awarded millions, stay in the same homes they originally said they wanted to leave, and then sue again.

The Diamond community wasn't so disingenuous as to hatch a plan to bank millions of dollars from a jury of its peers, stay right there in the shade of the chemical plant, and then shoot for the legal lottery again. Their nights had been too long to play such foolhardy games, and their families too sick. But the image of plaintiffs seeking riches from venerable Shell had been danced around the courthouse the past two weeks, and enough of it stuck to make a difference.

"Shell knew that our goal was to get people away from there. And perhaps we didn't do a good enough job of conveying that to the jury," Pendley reasoned.

Some of Margie's neighbors broke into tears, and one said, for all to hear: "All of this, for nothing?" Their four-year legal quest now abruptly behind them, the Diamond residents stood not one step closer to their ultimate goal of new homes away from industry. They were deflated, and some thought back to the dawn of the struggle, when a few of their neighbors right there in Diamond said it was a fools' game to think you could take down mighty Shell. Maybe the naysayers were right, after all. Who ever told Diamond it could outfox Shell Chemical?

Some, like Gaynel Johnson, kept calm, trying to make sense of the stinging defeat while plotting the next move. Officials from Xavier University were also there at the courthouse, and they talked with Diamond activists about ways to keep the fight going.

As the residents gathered outside the courtroom to take stock of the loss, Allen Myles walked briskly past, straight to an elevator, and left. He did not say a word.

Wilma Subra was at another meeting when she learned that the community had lost the case. She was stunned. The evidence was ample, she knew, that the community should get away from the plant. Common sense said it. Science said it. Even the law in Louisiana now said it. You have to be two thousand feet, not twenty-five, from a chemical plant.

Margie looked toward her father, Theodore, who had attended the trial each day, urging her on, and then she too walked from the courtroom.

The local media caught up with her, wanting to know what the community's chief organizer had to say now that the fight had ended.

Margie peered into the cameras, and she felt a calmness sister Naomi had experienced so often. She looked straight at the reporter. "This is not going to stop. We live in fear."

It's only the beginning, she thought. The whole world's going to know.

Ten

THEODORE
EUGENE

Not long after the courtroom loss, Margie and her father sat for a visit. Theodore had witnessed each day of the trial, and he saw what Margie had seen, and now he wanted to know how deeply the loss had affected his daughter, and whether it had withered her strength or inspired her anew.

"Dad, I just don't understand," she confided. "The lawyers had all the evidence they needed."

Father and daughter looked at each other across the room at 26 Washington Street. "Sometimes you have to be stronger than you think you are," Theodore said, and he asked Margie how she felt. Was she going to give up?

"I'm not going to stop. I'm just not going to stop."

"Make sure you gain all the knowledge you can," her dad said.

With those words, their brief chat came to a close with Margie plotting new ways to make things right.

Margie renewed her quest for knowledge, reading about government agencies from the DEQ to the EPA. She began making contacts outside Norco, getting the ear of environmental activists with a track record of challenging industry, and she gathered more brochures on chemicals, pollution, and the environment. She took her place as an activist speaking before government committees overseeing the Chemical Corridor, and as she pressed forth, she did so with the certainty that the struggle would be won only through community activism, not courts of law.

On Thanksgiving Day 1997, the family gathered at a cousin's house, and the atmosphere around the bountiful table gave no clue that the community had just suffered a devastating courtroom defeat. The talk on this day was of thanks, as the holiday inspired, and Uncle Brother looked around at his nieces and nephews, cousins, wife, daughter, grandchildren, and great-grandkids, and the quiet rock of the neighborhood beamed.

"My dad was an active person as far as moving was concerned, always was, and in his latter days perhaps his health got a little weak and he developed Parkinson's. Because of his determination, it didn't stop him, he didn't have the extreme shakes. He still drove his truck and planted his garden in the backyard. His relaxing time was sitting down after he planted a row of tomatoes. He'd sit down and rest in his garden."

A few days after the Thanksgiving feast, Uncle Brother stepped in from his outdoor work, but this day dizziness took hold of him with such force he tumbled to the floor in the hallway of his home. Aunt Mabel had always carried a loud country yell, and seeing her husband lying on the floor, she let out a scream for Margie to come over from her trailer next door. Margie hustled down the three concrete steps in front of her trailer and the few feet over to the house. Other kin got wind of Theodore's fall, and soon the house was filled.

As family hovered over him, Theodore tried to put all at ease. "I'll be okay, I'll be okay. Just let me lie down awhile." Margie wasn't convinced and wanted an ambulance to rush him to the hospital. Theodore wouldn't go that first day. "I think if I rest awhile, I'll be okay," he said.

The next day, he did go to the hospital. Doctors concluded that Theodore Eugene had suffered a stroke, had lost the use of much of his left side, and needed to check into a nursing home to receive the full-time care his health now required.

"I'm a nursing home," Margie, the recent school retiree, chimed in. "He doesn't have to go."

She began taking classes on how to aid people with strokes, and how to help the wheelchair-bound navigate their new lives.

"It was a battle but it was a joy," she said. "I became his primary caretaker. Family would help him in the evenings. I gave him his bath, got him up, dressed him, and put him in his wheelchair. He would like to ride out in the yard to see his livestock. All of us would feed the chickens. He still was giving advice, very strong, never complained."

On Christmas Day 1997, Theodore donned a Santa hat as usual, and he found a comfortable chair in the living room from which to take in the view of his great-grandchildren opening their gifts under the glistening tree inside the family house, whose exterior had been painted a soft peach and white.

Caretaker Margie took joy in the quiet moments, as when she changed his socks, taking care to rub his feet with lotion before dressing him each day. Daddy was thankful for the foot rubs and let Margie know it, but the truth was she could see that his strength was withering. He suffered even more strokes, weakening his condition further.

His grandkids—Margie's daughters, Caprice and Ericker, and Naomi's three children—took care of Grandpa too, putting on his favorite television shows and sitting near his side. Talking to him was another matter. Theodore Eugene spoke so softly you could barely hear him.

By March of the new year, Theodore was strong no more, and the family had no choice but to place him in the hospital's care.

On April 1, 1998, Margie was walking to Daddy's room when Mother stopped her in the hallway.

"He left us," Mabel Eugene said.

Margie broke, yet as she wiped at the tears, she kept walking right toward and into the hospital room. She gathered a bottle of cream, carefully unrolled Daddy's socks, and began rubbing his feet, one last time.

Eleven

"MINOR INCONVENIENCES"

Soon the Diamond community's legal fight was officially extinguished.

Just days after the courtroom defeat, the Diamond legal team hurriedly filed court papers to overturn the judgment or gain a new trial, arguing that the evidence presented in court was "strongly and overwhelmingly in favor of plaintiffs." The jurors got it wrong, the lawyers said, and the judge should set it straight.

Shell lawyers fought them fiercely. After all, the Shell lawyers argued, the community had had its day in court, and a just verdict had been rendered.

"The testimony as a whole supports the position that the minor inconveniences do not rise to the level of a legal nuisance entitling the plaintiffs to the relief requested," Shell Oil lawyers Mark A. Marino and Charles M. Raymond wrote, shooing away any thoughts of a new trial or new verdict. Judge Robert A. Chaisson agreed, at least with the larger issue that there would be no new trial.

Margie and her neighbors heard the words, and they stopped in their tracks—"Minor inconveniences?"

Three more such "minor inconveniences" occurred in Diamond shortly after the jury decreed that life in the community was not so much as a legal nuisance.

Three months after the jury verdict, state inspectors unearthed problems at the same Shell Chemical plant that had just been given high marks by courtroom jurors. An inspection by Louisiana's Department of Environmental Quality in December 1997 found that the company had failed to mark containers accumulating hazardous waste with the words "hazardous waste," and that it had failed to keep containers holding these hazards closed during storage, as required. More than a decade earlier, the DEQ had cited Shell for problems handling hazardous waste, the company had offered reassurances that all was well, and the issue had gone away. Now here was history repeating itself. Shell, once again, was directed to clean up its procedures in Diamond. It vowed to do just that, and the issue faded once more.

Then at six in the evening on February 10, 1998, a storage tank at the chemical plant became so overpressurized that the tank's roof shot off. As required, Shell filed an official "release report" after the incident. The company neglected to note one detail—that the massive tank roof had catapulted into the air and landed on the Diamond community's playground on Washington Street. The roof was so heavily girded that it would easily have killed anyone in its path. Miraculously, no one was there when it crashed onto the playground.

Shell's view was that the incident caused a "minor" vapor release of allyl chloride, a chemical that can cause shortness of breath or medical emergencies in high concentrations. Shell said the exposure was contained on-site, and was not serious enough to exceed Louisiana clean-air standards. Yes, part of a tank handrail did land off-site, but no one was hurt, no roads were closed, no fires occurred, the company said.

Another inconvenience visited Diamond on Mother's Day 1998.

That Sunday afternoon, the residents were outside barbecuing or inside hosting family visiting from out of state, and with flowers abounding on this holiday and everyone back from church, a big block party was in full swing. Suddenly a vapor cloud spewed from the plant and drifted over their heads, interrupting the celebration with coughing, sneezing, and runny noses. Everywhere, it seemed, a cloud of white powder blanketed them.

"All of a sudden all this white stuff started coming like smoke. Then all of a sudden you could just smell stuff coming and everybody started running," Sandra Campbell, a Diamond resident for years, later said in a story recorded by the community's advocates. "And all the babies in the playpens outside? I had a yard full of people out there and all that food just went to waste and they never paid me for that food." Shell told her she needed a receipt to prove how much she'd spent.

Gaynel Johnson, by now living on Diamond Road, was among those getting ready to enjoy a Mother's Day barbecue feast with family. Gaynel had made more trips to the hospital than she cared to recall, and a consequence of her sickness was that her throat could pick up something foul in the air before others got wind of it. On this afternoon she looked to the air and then took to the phones, calling her neighbors and urging everyone to dial Shell to find out just what was going on. Now more organized, the residents were trained to keep careful notes, writing down whom they called and what the Shell official said.

With not even Mother's Day a safe harbor, the residents turned to Randal Gaines, a lawyer from the town of LaPlace, to help. Pendley and Myles were no longer involved in the Diamond community.

Gaines is tall, wiry, and athletic looking, in the shape you'd expect of a lieutenant colonel in the National Guard who served as a patient administration officer for a MASH unit in Operation Desert Storm. He was born in southern Louisiana and worked as a tax attorney for the IRS in Texas before his wife's medical schooling at LSU resettled them in the Bayou State in the late 1980s. Gaines carries himself with the precision of the monogrammed glass holders he brings for visitors' water glasses, but he knows firsthand of racial injustice.

One of his civil rights cases involved his hometown of Lutcher, Louisiana. The town was majority white, though not by a large margin. Few, if any, blacks were able to win an election, as the majority white population typically voted in whites. Blacks would be elected only if they were on the slate approved by the whites, so in essence, the black community could not choose its own representative. Gaines filed suit against his hometown, keeping a promise to himself to make it the first legal case he would bring when he returned home. Once the suit was filed, Lutcher's leaders voluntarily sat down and reapportioned the town, creating two black districts, two white, and one at-large. The council now mirrors the town, which was his goal.

Another case targeted a second town in St. James Parish, Gramercy, population 2,412, where nearly one in five residents were black, but where at-large voting practically ensured that an all-white political body would be elected. This 1993 lawsuit, filed by Gaines and two other attorneys in federal court in the Eastern District of Louisiana, alleged that the deck was illegally stacked against the town's black voters, an arrangement that was both unconstitutional and a violation of the U.S. Voting Rights Act of 1965. The suit never went to trial, as after depositions had begun to be taken, the town's aldermen agreed to redistrict the town, creating a majority black voting district.

After a spill on Mother's Day prompted Margie and her Diamond neighbors to enlist his help, Gaines never had to file suit. An insurance company for a Shell vendor settled for nearly $400,000, spread among some seven hundred plaintiffs, with Gaines's firm and a fellow lawyer pocketing a 30 percent commission. The community learned that the vendor was delivering lime, a cleanup material, which spilled and got into the air like powder, producing a cloud—not toxic, but a bitter reminder of a life spent beside a chemical plant.

Shell explained that a transfer line from a tank truck unloading lime into its chemical storage silo had developed a one-and-three-quarter-inch hole that day, and that the northwesterly wind carried it off-site. "No regulatory reportable quantities were exceeded," the company declared, in a summation that meant little to the families whose Mother's Day had been ruined.

In the time he was involved with the matter, Gaines could see decades of built-up frustration boil over. "This is not the first time we've been exposed," the residents told Gaines. "We deal with this every day." He sensed they were on edge, and that the community wanted to press the case so that Shell would finally take their concerns seriously, just as Gaines had pressed voting-rights suits to get the attention of communities still governing with voting systems from the past. Far from retreating quietly after the courtroom loss, the Diamond community had decided instead to continue questioning Shell.

The lime spill was further evidence that the people of Diamond needed out, and Gaines was taken by their leader's resolve. Gaines and Margie became so close that she would later babysit for his children. He saw that she was feisty, energetic, and passionate about the environment. He also saw that she was different from some of her neighbors, who would encounter a mishap and simply want to get past it. Her passion, Gaines saw, did not cross the line to bitter hostility. Here was a community activist who was also a diplomat, and Gaines believed that if anyone could move the giant across the way, it was Margie.

Unbeknownst to her, the battle with Shell was making a difference, if only behind the scenes. After the courtroom victory, Shell's public face remained every bit that of the big company in the company town. The victorious lawsuit had vindicated it, as Shell surely knew it would. Privately, some at Shell admitted that Margie's rabble-rousing had forced the company to rethink its way of doing business, even if just a little. One day not long after the jury verdict, a Shell boss pulled aside Margie's daughter, Caprice, the lab technician. "Even though your mother's case was lost, she's forcing us to improve air quality," the boss confided to her. "She may have lost the case, but we're upgrading the facility."

Yet even with the books now closed on this latest nuisance to arrive at Diamond's door, the community remained in the same rut it had been in for decades. It seemed every time there was a flare-up, the lawyers came knocking and industry paid out a small sum. Margie knew settling these small incidents took the community no further in its quest for ultimate resolution—relocation from the plant.

In March 1998 came the official word from Shell, a simple, stinging sentence aimed toward the Diamond residents still holding out hope.

"Relocation," the company said, "is not an option."

Outside of Louisiana, Shell was not escaping court cases with the ease with which it beat back Diamond's lawsuit. On September 9, 1998, the U.S. government and state of Illinois announced the conclusion of a federal court civil case brought against Shell Oil Company, Shell Wood River Refining Company, and others, accusing Shell of committing hundreds of environmental violations at its oil refinery on the banks of the Mississippi River in Illinois, not far from St. Louis.

Standing under the famous Gateway Arch in St. Louis, Attorney General Janet Reno and federal officials announced that Shell would pay a $1.5 million fine for its neglect. The fine was small for such a large company, but the government believed the case showed it would not tolerate polluters, and the headlines about the crackdown did not enhance Shell's image.

The long list of environmental failings included illegal levels of sulfur-dioxide and hydrogen-sulfide air emissions, plus violations of benzene emission standards, Illinois water regulations, and solid-waste labeling requirements. Further, Shell failed to timely report emissions of "extremely hazardous" substances including ammonia and chlorine.

Beyond the fine, Shell's penance included signing an agreement requiring it to comply with all environmental laws and perform cleanup projects worth $10 million. Shell further promised to expand water quality and wildlife protection by buying $500,000 worth of land adjacent to the Mississippi River, and then handing ownership to the state of Illinois, which would preserve its wetlands, water, and wildlife. Another project would reduce air emissions of sulfur dioxide and nitrogen oxide.

"In settling this case, the federal government has followed the basic principle that polluters will be required to pay for and correct the damage they cause, as well as prevent future damage," said Steve Herman, the EPA's assistant administrator for enforcement and compliance assurance.

Publicly, Shell downplayed the government's case, saying that many of the allegations had long been corrected, but that it signed the consent decree to avoid the cost of litigation. To Attorney General Reno, the case was no small matter. "The Mississippi River is a part of our national heritage. We have a responsibility to restore and protect it not just for this generation, but also for all of those to come. To those who think that they can get away with illegally polluting our river, we say this: we will work together at all levels of government to find you, prosecute you and make you clean up the mess you've made. You could even go to prison."

Back in Norco, complaints about environmental damage had elicited no such passion from government, just the occasional notice that Shell should clean up its act. Politicians surely weren't clamoring to punish Shell, one of the Chemical Corridor's major employers. The Shell official's confidential words to Caprice notwithstanding, few in the official corridors of power were listening to what Margie Richard and her neighbors in Diamond had to say.

In the spring of 1998, eager to get the ear of anyone she felt could help the neighborhood's cause, Margie went to an environmental law conference at nearby Tulane University, a casual setting where lawyers and activists spoke about the challenges they encountered in helping communities smothered by pollution. As the discussion evolved into an open Q&A session, Margie raised her hand, stood up, and spoke about life in Diamond, not thirty minutes from this gathering of lawyers and academics. She told of the funerals she had attended and the hospital visits she had made, stressing throughout her community's prime wish of relocation.

Sitting in that conference was Monique Harden, an attorney with the Earthjustice Legal Defense Fund of New Orleans, a nonprofit environmental law firm. Harden was struck by this woman wearing a green print dress and gold wire-rimmed glasses, her hair up in a bun, who interjected a passionate dignity into the relaxed setting. Afterward, Harden could recall few of the other attendees that day, but she could not erase from her mind the story Margie Richard had told, so Harden decided to see for herself what this woman was talking about.

A few months later, she drove to Margie's trailer for an evening sit-down with residents where, over soft drinks and snacks, she would share insights to help the community achieve its goals. As she stepped out of her car, Harden was greeted by plant smoke so thick she thought she had somehow been transported to a smoke-infested rock concert, or to the foggy streets of London. When she entered Margie's trailer, the dozen residents gathered there told her this was simply what it was like with a chemical plant as your neighbor.

The meeting began with a prayer, and then the lawyer and community made acquaintance, residents telling Harden about the asthma and other breathing problems that infected every street of Diamond. One resident told her that many in the Diamond neighborhood slept "ready robed"—with their clothes on, so they could quickly flee come the next explosion. The Earthjustice lawyer resolved to help these people, but she knew a solution would not come easily.

Diamond remained isolated, every bit as divided from the rest of Norco as it had always been, the thick grove of trees that separated white and black communities still serving as a literal divider not only of the physical communities but also of the opinions each side held about the benefits of life along the Chemical Corridor.

Another long-term legal matter continued to expose that divide—the murder conviction of Gary Tyler, the black teenager sentenced to life in prison for the killing of a white boy amid the school desegregation tension of 1974. By now, two decades later, defense advocates had amassed evidence they believed proved Tyler did not commit murder but was wasting away in prison because of brutal police tactics and questionable trial testimony.

Their evidence was detailed in an Amnesty International report. The bus where Tyler had been sitting, and from where he allegedly fired the gun, had been searched literally inside out over three hours in the wake of the shooting. Police found no weapon. Later, after a female student sitting next to Tyler told police he had indeed fired the bullet, authorities suddenly found a .45 automatic stuffed inside the seat. "The seat had been previously searched, shaken, and turned upside down several times and nothing had been found," Amnesty International wrote.

This same child fingered Tyler on the stand. After the trial, she recanted, as did other witnesses, who said the police had coerced their statements with strong-arm tactics. The gun? It had been traced back to a police firing range used by St. Charles Parish, from where it had allegedly been stolen. Toss in questions about the legal representation provided Tyler, and his conviction by an all-white jury, and Amnesty International and others were convinced justice had been abused. The U.S. Court of Appeals added credence to these concerns, once writing that Tyler's trial had been "fundamentally unfair."

Three times, the Louisiana Board of Pardons voted to reduce Tyler's sentence, which would have granted his freedom after twenty years. The family of Timothy Weber fought the release every time. "I'm going to keep coming back until I'm

old and gray," Leah Weber told reporters one afternoon, convinced that he was the killer of her child. Three times, Louisiana governors refused to free Tyler, the last of which had
been in 1995.

To the black community, this was proof that two scales of
justice existed for southern Louisiana, one for black residents
and one for whites. The battle over big industry was yet one
more case study of a region living in two distinct worlds.

In October 1998, Shell commissioned a survey of people in
Norco specifically, and of the parish of St. Charles as a whole,
dialing 315 residents for their opinions on industry's role in
the town. The company conducted the research in conjunc-
tion with Loyola University in New Orleans and community
groups.

Nearly three of every ten people questioned worked in the
chemical industry or had a member of their household who
did, and more than three-fourths of the respondents were
white. The sample was a mirror of the town—mostly white,
and solidly proindustry. Their livelihoods, after all, depended
on it.

Not surprisingly, whites had a more favorable view of the
petrochemical industry than blacks, and blacks were less trust-
ing and appreciative of the benefits of Big Oil. The reason was
simple. Whites were far more likely to be on the industry payroll, and they trusted the companies paying their bills. Blacks,
nearly shut out from industry, did not. "Nearly half of the respondents thought the amount of pollution from petrochemical plants was staying the same, but black respondents were
more likely to think emissions were increasing compared to
white respondents," the findings showed.

Shortly after, the study organizers commissioned a two-hour focus group meeting in Diamond, at the Good Hope Missionary Baptist Church on East Street, and they heard residents express strong concern over the lack of job opportunities and a plainspoken fear that industry was not communicating with them. Yes, the residents appreciated what Shell did in the local schools, but worry over emergency response, health risks from pollution, and nuisance issues topped the concerns.

"The group feels invisible to . . . Shell," the report found. "Generally, the closer citizens live to the fenceline, the more they worry."

These were only words. Margie and her neighbors were champing at the bit for action.

As *Richards v. Shell* found its way to the court archives, no longer an active matter of law, there was no inkling that potential disaster would pay a visit once again to Diamond.

Part
TWO

"HOW WILL I KNOW IF THERE IS AN EMERGENCY?"

If there is an incident which could endanger the Shell Norco Chemical plant area residents, *the St. Charles Parish Outdoor Alerting System will be sounded. You should seek shelter immediately.*

"WHAT SHOULD I DO IF I HEAR A WARNING SYSTEM SIREN?"

If you hear the St. Charles Parish Outdoor Alerting System, immediately go inside a house, building or vehicle. *Close all windows and doors and shut off your air conditioner or heating unit or any other ventilation system. If you have a fireplace you should also close the damper at this time. Cover any open areas, such as around doors and windows, with damp towels or sheets. If you should have any trouble breathing, place a wet cloth or towel over your nose and mouth and try to breathe in quick shallow breaths.* Most importantly, remain calm. *Wait for further information from either the St. Charles Department of Emergency Preparedness or Shell Norco Chemical. . . .*

"WHAT IF MY CHILDREN ARE IN SCHOOL?"

The teachers and administrative staff at your children's school have been trained to protect your children in the event of an emergency. *You should not attempt to pick your children up at their school.* . . .

"WHAT IF I CAN'T FIND SHELTER?"

An enclosure, in any form, is the best protection in a chemical emergency. If you can't find shelter, move so the wind is blowing from your left to your right or vice versa. Do not move with the wind blowing directly into your face or from behind you.

—Community Awareness and Emergency Response pamphlet, sponsored by the St. Charles Parish Department of Emergency Preparedness and Shell Chemical Norco

Twelve

NORCO
ELEMENTARY

On December 8, 1998, at just after eight on a Tuesday morning, Shell's how-to manual in event of chemical catastrophe became starkly relevant for the people of Diamond, their neighbors, and schoolchildren.

From the belly of the chemical plant came a rumble loud enough to be heard at Norco Elementary School, a half mile away. A highly flammable chemical began to overheat inside a five-thousand-gallon Shell tank in the plant's batch resins unit, where epoxy resins are manufactured for use in circuit boards through a process that heats resins and methyl ethyl ketone, or MEK, a paint solvent, to form another resin. This morning, the material's thickness increased far more rapidly than normal, indicating an uncontrolled reaction was afoot, and bringing the frightening prospect that the pressure would rise to such intensity that the tank would rupture.

By 8:15 A.M., the event was referred to by Shell as "an unusual event," and an on-site emergency response team went to work. As the pressure cooker heated, a more urgent alert went out at 8:27 A.M., letting parish officials know a potential catastrophe was brewing, and emergency personnel needed to be dispatched throughout town.

Shell employees sweated furiously inside the plant, laboring to somehow lower the tank's temperature. Managers feared the tank was going to explode.

The incident commander became so concerned that flying debris from a ruptured tank could reach the elementary school, he had the school sealed shut, with children inside, to avoid that horrifying scenario. Teachers dispensed water and towels, instructing the now-quarantined children to hurriedly cover their noses.

Younger kids en route to classes were turned away and sent to school in Destrehan, the state police were dispatched, and Shell personnel, with radios in hand, fanned out to nearby schools. As the media caught wind that Norco was once again in a panic, Shell employees scurried to the four-street Diamond community, banging on doors to tell neighbors to seal their houses and stay inside and, by whatever means, to cover their faces and be prepared to take quick, shallow breaths.

Seventeen times over the next three hours, Shell personnel dialed the red hotline phone at the parish emergency preparedness office. To alert the town to what was potentially afoot, 730 messages were delivered in just fifteen minutes, and staffers with Louisiana's Department of Environmental Quality rushed to the scene.

Three and one-half hours later, the alert was called off.

The chemical bomb in the making had been contained, and frantic parents rushed into Norco Elementary to pull their children out. Students, many still trembling, ran to their parents' cars and made their way back to their homes, some still covering their faces as they left the school.

Teachers and parents of the children shuttered inside the elementary school could not help but wonder, What if the plant had blown?

Life in Diamond may not have been a nuisance in the eyes of Shell and a St. Charles Parish jury, but with frantic school shutdowns, houses locked down, Mother's Day lime spills, and the specter of Shell tanks hurtling in the air, a "nuisance" seemed like the least of the Diamond inhabitants' worries.

Gaynel Johnson had grandchildren and great-nieces attending Norco Elementary, and some of her grandkids, like her, were asthmatic and needed breathing pumps to get through the day. That day, her grandchildren had been routed to Destrehan, and when Gaynel finally did see them, she hugged them tight.

As Margie ventured home from Bible studies class that December day, she witnessed the scene outside her home and thought of her ailing mother. Oh Lord, she thought. Not again.

Theodore Eugene had passed away a few months before, and now Margie was Mother's rock. She stepped out of her car and into the family's Washington Street home to comfort her mother, but as she walked the few brief steps to the house, the air was filled with a mist.

"Mom, what happened?" Margie asked. When she got word that yet another mishap had paid visit to Diamond, she told Mabel, "Mom, get your clothes. Let's get out of here."

Looking at her mother and smelling the foul air once again, Margie knew instantly that this day would mark a turning point in the fight with Shell. "I felt like David and Goliath. I felt like Esther," she said. "Only God kept it from exploding."

Authorities assured the masses that all was fine, the seventeen telephone calls to the emergency hotline notwithstanding, and Shell insisted that the disaster had evaporated and that everyone should just go about their day. The company said its air sampling detected zero levels of MEK in the community. MEK is a flammable liquid that can trigger skin rashes and burning on contact and cause severe damage to the eyes. Breathing too much MEK can result in heavy coughs and wheezing, and overexposure can bring dizziness, blurred vision, and nausea. Repeated high exposure can affect the nervous system and the brain. Shell insisted no such toxins, not even in a small amount, could be found in the air.

If that was the case, Margie wanted to know, why was a chalky cloud now rising from the same plant, and why were her eyes moist with the sting? The answer, the community would later learn after pressing the matter, was that a second toxic discharge had come from the Shell Chemical plant that very day, this one involving a release of hydrochloric acid, HCl.

At ten minutes past noon, precisely thirty minutes after Shell and parish officials declared an all clear on the tank that did not explode, an overfilled railroad tank car spilled four hundred pounds of material, and a visible plume of hydrochloric acid wafted over the maintenance building and into the community. The Diamond residents were told little, if anything, about this second scare that day.

Margie dialed the Louisiana DEQ, the state agency in charge of regulating the Chemical Corridor, in search of an answer and a salve, at least in the short term, for the community's newest bitter taste.

The DEQ put her on hold.

Seconds passed, and the phone line felt lifeless, every bit as inept, Margie felt, as the government's response to the community's pleas had been these long years. She hung up.

She picked up and dialed again, but this time it was not another call to the red tape that was Louisiana's government. The second call went to Denny Larson, a California environmental activist Margie had met during Larson's grassroots travels. At the time, Larson happened to be in Baton Rouge for environmental meetings headed by the EPA's National Environmental Justice Advisory Council, or NEJAC, sessions that were exploring issues of concern along the Chemical Corridor.

In fact, the environmental contingent had toured Diamond just two days earlier, on Sunday, in a visit Earthjustice lawyer Monique Harden had helped arrange. Four buses traveled from Baton Rouge to Diamond, and Harden made sure the drivers drove slowly through the community, so the EPA officials could see just how close the homes were to the plant and take stock of the limited ways in and out of the neighborhood in the event of a catastrophe. On Washington Street one of the buses nearly tipped over into a sewage ditch, so the others used extra caution while navigating the narrow roads. From her bus, Harden was struck by how clear the air was that morning: Shell had known the tour was coming. The lawyer thought back to the unannounced visit she had made to Mar-

gie's trailer just a few months earlier, when the air was thick
with releases from the plant. When EPA brass wasn't in town
for a sightseeing tour, she surmised, the sky was not so pretty.

On Monday, December 7, day one of the conference, Margie Richard took to the podium at night and spoke of the suffering in her community. Samuel Coleman, the EPA's
Dallas-based director of compliance and enforcement for the
southern United States, sat spellbound, as he had three years
earlier when he'd heard Margie speak at a conference in San
Antonio. "Margie said, 'We get these scares, the children are
afraid, parents are running for their lives, you can't get any
information from the company,'" Coleman said.

Don Baker, Shell's liaison with Diamond, pulled Coleman
aside that night in Baton Rouge. "It's not like that," Baker insisted.

Now, the day after Margie's speech and two days following
the bus tour, reality had hit again, and Margie turned to Larson. They had first met in 1995, before the community's lawsuit
had gone to trial, the acquaintance coming through a mutual
friend, chemist Wilma Subra, who had worked alongside Larson studying the refinery corridors of Texas and Louisiana,
home to the two largest stretches of petrochemical plants in
the United States.

Their first meeting had also been Larson's introduction to
Diamond. As he settled into her trailer home that day, Margie
asked, "What would you like to see?"

"I'd like you to walk me around the neighborhood and tell
me the story of what is happening here."

They stepped into the street and began a long day's journey, with Margie pointing to the spot on Washington Street

where two neighbors had been killed twenty-two years earlier, and noting house after house where sick neighbors resided still. She told Larson how her sister died of a rare illness, how the blast of 1988 was so fearsome it sent debris hurtling onto the elementary school, and how daily life in Diamond was filled with uncertainty.

"We want to be relocated and we want a fair price," Margie said. "We've been fighting for a long time, and we'll fight as long as it takes."

As Larson listened while they walked, he noticed the occasional vacant lot between homes in Diamond, and it occurred to him that Shell had already been buying out properties piecemeal. The recognition lit a torch for an activist who had long studied the Chemical Corridor of southern Louisiana, but who was now getting his first ground-floor view of Norco. Diamond was the poster child for a community that should be freed from industry. It wasn't just that homes abutted industry, a problem throughout the region. In Norco, there had been death in the street, scares at the school, and unexplained illnesses in home after home. "It was very clear that people were in harm's way," he said. "It wasn't just a theory."

For years Margie had been trying to solve the community's ills by getting Shell's attention in Norco, but in Norco, Diamond's voice carried little weight. Now, walking with Larson in that first get-together in 1995, she told him she and her neighbors had to reach a larger audience, outside town, and maybe outside the country, in order to get the attention of the plant next door. "We want to take the fight beyond Norco," Margie said.

Larson, himself an environmental globetrotter, observed

how challenging life by the plant must have been, and he agreed to help spread Margie's message. He snapped some photographs and put Margie's story up on the Web site of the National Oil Refinery Action Network, whose goal was to help small communities win environmental struggles.

He helped her take a spot on an EPA-appointed federal panel looking into the refinery industry, the august Petroleum Refining Sector Subcommittee, whose members included the likes of the vice president of ConocoPhillips, the head of Marathon Ashland, and others from the oil industry, plus top EPA people. Margie's voice of down-home activism injected fresh energy into the committee of white-collar brass and career bureaucrats. She was now rubbing shoulders with people whose view was bigger than Norco, just as she'd told Larson she needed to do.

"We had to march on," she explained. "When we lost the court case, I knew we had to go beyond Norco. I knew we had to have technical facts. I kept having those nightmares about Norco exploding."

She reenergized her neighbors, and they changed the name of their group. Now it would be called the Concerned Citizens of Norco. This time, Margie knew, the community activism had to carve the path to lead them out of Diamond. The only way to finally effect change was for the people to do the work themselves.

In an attempt to achieve tangible results, she had long ago created a systematic program to take the campaign to the next level. With a list in hand of Shell executives, she started dialing. If a Shell executive didn't answer, Margie, instead of going down the list to that person's assistant, rang that executive's boss. If the boss didn't answer, she called the next executive. In each case, the secretary took her name and number, but Margie rarely heard back in any meaningful way. She kept at it nevertheless, believing someone would want to hear what her community had to say.

Denny Larson was drawn to the Diamond community's cause, and to Margie's. With his goatee, sandy-colored hair, and youthful mug, he appears one-half beatnik, one-half college professor. He favors faded jeans, sandals, baseball caps, and black T-shirts emblazoned with slogans—"Educate, Agitate, Organize"—in bright colors. He'll don a blazer, but nothing too fancy. At first glance, Larson comes across as the laid-back Montanan that he is, at ease talking to intimate grassroots groups, but he's a well-versed and vigorous agitator on behalf of small communities whose voices are rarely heard.

His campaign is global, and his e-mail inbox is too. One day it contained 3,928 messages, all of them read. His eyes are tired around the edges, perhaps from his world travels and computer screen, but he has a clear idea of what communities should do in their battle with big industry: Keep logs of bad smells, "stink diaries" some call them; swab the crud off your car and preserve it for testing; and take air samples, plenty of air samples.

At the moment Margie rang him during the elementary school scare, Larson was delivering to communities his own pollution-fighting gadget, originally made famous by the lawyer who employed the celebrated environmental renegade Erin Brockovich. It was a bucket residents could use to test the air in their neighborhood for toxins and other health hazards. Larson had been in Contra Costa County, California, as an environmental activist when Brockovich and attorney Ed Masry fought a polluting refinery there, and he got to know them as the crisis prompted Masry to search for a bucket that could test the air. Masry turned to engineers, who invented just such a device.

Larson worked side by side with Masry in California, and delivers an uncanny impersonation of the late lawyer made famous in Albert Finney's Oscar-nominated performance in the film *Erin Brockovich.* "I need something to test the air!" Masry would bellow.

There, in Contra Costa, a Northern California community had encountered a toxic cloud of chemicals from a nearby refinery, a bitter visitor that stayed sixteen days in the middle of summer, when many people leave their windows open at night. At first the company didn't tell the people or public officials, and it went about operating its refinery as if nothing had happened. "By the sixteenth day, the town was covered in a brown goo," Larson tells colleagues and communities he comes in contact with. The residents soon found the source of the goo, and by day sixteen, the press was all over the story.

Matters got worse some two months later when a subsequent release, of a different chemical, forced a local school to be evacuated and dozens of children to be dispatched to the hospital. Yet another spill made Masry and Brockovich sick.

When Masry and Brockovich came to sue, they couldn't initially prove a release had occurred, because no one had monitored the air. This problem inspired the bucket testing. Masry's idea was to give monitoring devices to their clients, the residents, since the government and industry couldn't be trusted to be forthcoming about what was happening. The community Masry and others represented later won an $80 million lawsuit settlement. The community had been sickened, but it took the people, not the regulators or industry, to prove it. Activists, including Larson, soon asked Contra Costa officials to sanction the Bucket Brigade's work, and the county later became a partner in helping to fund the testing of air.

Contra Costa County was not the case that brought Brockovich fame; the better-known crusade—proving that the Pacific Gas & Electric Company had poisoned the water in Hinkley, a desert town in Southern California—had just started at that time. Though the PG&E case generated a record $333 million settlement and a popular movie, the Contra Costa battle may ultimately prove to have a more lasting impact on grassroots struggles because of the buckets that were created to test the air.

Afterward, Larson got Masry and Brockovich to donate buckets that had been used in that fight, and from there Larson expanded the testing of air to other communities, creating the Bucket Brigade concept. In no time Larson was hearing from people across the country asking how they could get their hands on those buckets.

He worked to perfect the model, and it evolved into an ingenious contraption comprising a five-gallon plastic bucket with an airtight lid, a patented, specialized bag, and a battery-operated device that draws air in. He even wrote a how-to manual.

"This is just like a mechanical lung," Larson tells his community disciples. "This lung is going to breathe in those same odors. In the heat of getting gassed by one of these companies, you can run out and take an air sample." He says it so matter-of-factly you forget for a moment he's not talking about some school assignment, but rather a mission to catch potential toxins in the air.

Larson comes to a town only by invitation, and only if there's a band of hard-core activists ready to work. What they're doing is not a research study, Larson likes to say, but a battle waged in sometimes perilous conditions. Residents must step out of their homes with the buckets, get the apparatus going to soak up the air, then hustle to FedEx so the mechanical lung can be shipped to a lab in Canoga Park, California, for analysis. This procedure, Larson says, allows small community groups to conduct the type of testing that companies or regulators often don't, and in the process, they get their hands on reports detailing levels of contamination that had for years, or decades, been ignored.

"These agencies want to do the sampling at the wrong location at the wrong time," he tells his trainees, and he urges action at the very moment the air is most foul, not hours later. "We do the sampling at the hot spots in the fenceline."

His bucket works, as its EPA certification shows.

When Larson visits communities across the globe, he reminds them not only to test the air but also to keep logs of smells from big industry, to take pictures, and to roll the video. "Nobody else has that kind of information. The agencies and the companies are not going to do it until you do," he says. The Bucket Brigade has operated in twenty-five to thirty spots in the United States, and fourteen others around the world. "You've turned the tables on the agency and the company," he says in his voyages across the globe. "And now you're in charge."

The Bucket Brigade and other forms of activism recall the pioneer days, when everyone chipped in to battle a fire at a neighbor's house, using their own hands to do it. "We are joining together and working together to put out the fire," Larson says. The image is where he drew the name Bucket Brigade. In many of these communities, Larson reasons, there literally is a fire burning.

So it was on that December day in 1998. The two events—the near disaster and the acid release—had happened hours earlier, but Margie told him: "We still smell it."

Larson hopped into his rental car and drove straight from the Hilton Hotel conference site in Baton Rouge to Margie's trailer in Norco, a seventy-minute trip, then stepped outside and set about taking air samples from the plant that was choking the community. "Take us to the smell," he said.

They found a spot at Washington and First streets, and at 4:41 P.M. on December 8, 1998, residents took the first Bucket Brigade sample of air in Diamond. It was so late in the day that they had to rush to make the FedEx deadline to get it to the lab in California, Columbia Analytical Services.

Larson called the lab to stress the urgency of their work. Some samples can take more than a week to be evaluated and returned. Larson felt the Diamond inhabitants didn't have the luxury of time. Not only was there an environmental gathering in Baton Rouge that week, but another event was coming the next weekend in New Orleans, hosted by the Deep South Center for Environmental Justice at Xavier University. National environmental officials would be there, and the press too. "It was a dream scenario for us," Larson explained. "There was built-in media, a built-in event, a built-in *everything*."

With this captive audience fortuitously drawn to southern Louisiana at that very moment, he pressed the California lab. "There's been a chemical spill. The people can't wait ten days," he said. "Can you do it?"

The answer was yes.

When the results came back, Larson quickly scanned the report for MEK, a dangerous toxin the activists believed had been released into the air, despite Shell's statement to the contrary. The lab report in his hand now did not list MEK, and for a moment Larson was perplexed. Then he checked again and saw that the report had listed the toxin under its other name, 2-butanone, and he knew the community was onto something.

Officially, Shell would later describe the dual events that day by saying, "Plant officials and local authorities determined that there was no release" during the elementary-school scare, and that the second mishap, involving hydrochloric acid, "did not cross the fence line before the cloud of HCl dissipated in less than two minutes."

Translation: No big deal.

"They said, to the public, to the press, none of this had gotten out," Larson told the residents. "Shell came out and handed out a piece of paper, 'Nothing happened.'"

Now his Bucket Brigade had proof that something had indeed happened, and his reaction was simply, "Holy shit! This was the first *gotcha* moment for the Bucket Brigade" in Norco. "It was the start of the first snowball effect that never let up.... We knew we had them."

With the ammunition in hand, Larson next relied on Wilma Subra to put the findings into context: to tell the community what was in the air and to specify what dangers these chemicals could bring.

Subra's report documented that MEK was indeed in the air, and that the chemical "irritates nose, throat, skin and eyes," but the more important point was this: Shell had denied such toxins were released. And, as Subra's report made clear, MEK was not the only one.

Larson scanned the results once more and called Margie Richard to let her know what that first bucket test had captured. Even with samples gathered hours after the events, the California lab had also found smaller amounts of toluene, which affects developmental and reproductive systems; acetone, which irritates the nose, throat, lungs, and eyes, and causes headaches; carbon disulfide, which irritates the nose, throat, and eyes; and a small hint of benzene, a carcinogen.

None was in an amount that exceeded Louisiana's ambient air standards. The big company portrayed this news as insignificant, saying the concentrations were so minor that they were harmless; but the results lit a torch in a community that just a year earlier had been told it had no case at all.

Larson considered the serendipitous series of events—the environmental conferences, the elementary-school scare, the testing of the air—and believed that Diamond's fortune was about to change. With national environmental groups and the press in town, people would start to take notice, and the regulators would now have to pay attention.

A day after the school scare Margie returned with Larson to Baton Rouge, scene of the national environmental conference, and she took to the microphone and recounted her telephone call of just a day earlier. "I picked up the phone, and I called the DEQ and was put on hold," Margie told the crowd. "I still haven't heard from them. It's the same old, same old."

Soon Larson and community organizers mobbed Samuel Coleman, the EPA compliance official based in Dallas, and barraged him with questions about the ongoing problems at a plant that continued to threaten the people of Diamond. Larson captured the scene on his video camera, and the EPA official, backed into a literal corner, acknowledged that the swelling group had a point.

"It's on the top shelf," the EPA man told those surrounding him in the hallway in Baton Rouge. "I have questions as to why these things happened, and I think these are legitimate questions."

Coleman knew Diamond's story well from his earlier visits. "I knew all of these people, I had been to their homes," he said. "I was confronted by a fairly large and angry group of people. They really wanted some answers. That's what I told them, 'We're going to get to the bottom of this, we're going to figure out what has happened.'"

Some months later the EPA hit Shell Chemical with a $27,750 fine for the mishap at the chemical plant on December 8, the biggest federal levy against the plant in nearly a decade. The fine, the EPA complaint said, was for Shell Chemical Company's violation of the federal Clean Air Act "by its failure to maintain and operate air pollution control equipment in a manner consistent with good air pollution control practice for minimizing emissions." Human error, the EPA concluded, led to the spill of 148 pounds of hydrochloric acid into the air that December day.

Coleman personally signed the government's official order. After the verdict in 1997, the Diamond community had been down. Most assumed its rabble-rousing was pretty much gone for good. Now Margie was grabbing the microphone, neighbors were lining up behind her, people outside Norco were paying attention, the community was gathering evidence with its own hands, and the government was taking notice.

The near explosion by the elementary school had lit the torch, and Diamond's grassroots movement had suddenly sprung back into action.

By now, Shell's chemical plant was taking hits from several corners. In January 1999, the Louisiana DEQ's Air Quality Division had issued a modest fine, of $7,000, against Shell for the May 1998 release of eighty pounds of allyl chloride vapor, a flammable liquid that can irritate the eyes and cause severe burning of the skin. Breathing allyl chloride can irritate the lungs, causing shortness of breath, and high exposures can lead to a medical emergency. A state inquiry concluded that the company's failure to properly handle equipment caused the release. For Shell, the fine was tiny—the company had reported gross revenues in 1997 of $32.2 billion—yet this penalty showed that the state was now paying more attention, just as the EPA was.

State records listed more than a dozen fines or citations brought against Shell in the decade from 1986 to 1996, covering a myriad of violations such as failing to submit required government emission reports and not maintaining pollution control equipment. Taken together, the small cases—some fines were as little as $4,750, and some inquiries resulted not in fines but compliance orders—added heft to Margie Richard's long-running argument that the plant was foul, and the residents should move from harm's way.

That same month, January 1999, the community and company agreed to meet to work through the turmoil of just one month earlier. Shell wanted residents to come to the plant, have their driver's licenses on hand so they could pass through security, and then head in, but the community activists argued they should meet on neutral ground, and the session was moved to a veterans' center nearby in Norco. Dozens of Diamond residents piled in, and so did a handful of white Shell workers. Damu Smith, an official with Greenpeace from Washington, D.C., opened by asking everyone in attendance to raise their hand if they wanted clean air.

Instantly a sea of black hands shot toward the sky. At the Shell table, one woman started to put her hand up but then, noticing that no one else at her table had, quickly put it down. Earthjustice lawyer Monique Harden glimpsed the sight and, in that small moment, witnessed just how powerful the company was in Norco.

In March 1999, Margie was a guest on *Making Contact*, an international radio program that was devoting a segment to environmental justice. Shell declined the program's offer to send a representative, so Margie took center stage.

"We're not making Norco look bad, as we've been accused of," she said. "If we don't tell them, how will they know? We're not doing it because we're out to get anybody. We're asking for help because we want our children to grow up healthy. . . . We were here first. The Shell Chemical plant came after."

One month later, Margie was invited to speak before the United Nations Commission on Human Rights in Geneva, Switzerland, as part of a small delegation of black environmental justice organizers and activists. Before she went, Margie gathered at home with her neighbors from the Concerned Citizens of Norco. "God," Gaynel said, looking upward, "cover her and protect her." In this meeting before her trip to Geneva, Margie confessed that she did not know exactly what she would say once she had the UN's attention, but little worry crossed her neighbors' minds.

Among those sitting side by side with her in Switzerland were Beverly Wright, director of Xavier University's Deep South Center for Environmental Justice in New Orleans, which had produced the sickness study in Diamond two years earlier; Harden of the Earthjustice Legal Defense Fund of New Orleans; other activists from Louisiana; and the International Human Rights Law Group from Washington, D.C.

Their goal was to raise awareness about environmental racism in Louisiana's petrochemical corridors. The Louisiana contingent was recognized as the first American delegation to bring the issue before the United Nations. On the table outside the conference the group passed out booklets for the attendees to read, titled "Louisiana Personal Stories"—eleven tales from the Bayou State of minority residents whose communities bordered on plants and pollution. There was Elodia's story from New Orleans, Florence's from Alsen, Minerva's from Convent, Albertha's from White Castle, and, among others, Margie's from Norco.

Sitting in a grand palace overlooking lush hills, Margie could feel the nerves wreaking havoc on her stomach. Stepping out of the main hall for a breath, she found a seat by herself. How in the world did I find myself in Geneva, Switzerland? What can I possibly teach the United Nations? she thought, visualizing the hall full of lawyers, doctors, and experts, people who'd never experienced life next to smoky industry.

Hey, I'm a nobody, she thought, and the doubt hit her so deeply she had to take a walk, by herself, outdoors. Margie thought back to her sister, Naomi. This is not for me, she reminded herself, and as she soaked in the beauty around her and thought of the snow-covered mountains she had seen earlier that day from her hotel room, a calm settled over her, and she said a silent prayer.

Margie walked back to the grand hall, stepped to the podium, and sat alongside her colleagues. That day, the UN Commission on Human Rights learned about the toxic emissions, deadly explosions, and illness in her neighborhood, the words coming from this slight woman with a lyrical voice wearing glasses, a colorful red headband, and a business suit.

"My name is Margie Eugene Richard. I am president of Concerned Citizens of Norco. My hometown is located in the southeastern section of Louisiana along the Mississippi River. In 1926, the Royal Dutch Shell Company purchased four hundred and sixty acres of the town called Sellers and began building its oil refinery. When Shell purchased the town of Sellers, which is now Norco, they displaced African-American families from one section to another.

"We are now surrounded by twenty-seven petrochemical and oil refineries and counting—refineries from which Norco received its name," she said. "Nearly everyone in the community suffers from health problems caused by industry pollution. The air is contaminated with bad odors from carcinogens, and benzene, toluene, sulfuric acid, ammonia, xylene, and propylene—runoff and dumping of toxic substance also pollute land and water.

"My sister died at the age of forty-three from an allergenic disease called sarcoidosis, a disease which affects [less than] one in one thousand people in the United States, yet in Norco there are at least five known cases in fewer than five hundred people of color. My youngest daughter and her son suffer from severe asthma; my mother has breathing problems and must use a breathing machine daily. Many of the residents suffer from sore muscles, cardiovascular diseases, liver, blood, and kidney [ailments]. Many die prematurely from poor health caused by pollution from toxic chemicals. Please indulge me while I share with you a few stories that embody some of our fears, because these tragedies can happen at any moment without notice."

She told the UN of the deadly explosions of 1973 and 1988, of a 1994 acid spill, and of the 1998 lime cloud and elementary school scares.

"Daily we smell foul odors, hear loud noises, and see blazing flares and black smoke, which emanates from those foul flares. The ongoing noisy operations and the endless traffic of huge trucks contribute to the discomfort of Norco citizens. We know that Shell and the U.S. government are responsible for the environmental racism in our community and other communities in the U.S. and many communities throughout the world. There must be an end to industry pollution and environmental racism.

"Even as U.S. citizens, we are not protected from environmental racism in the United States of America by our government. I would like to see justice in action that leads to an end to this struggle. Norco and many other communities of color across our nation suffer the same ills. We are not treated as citizens with equal rights according to U.S. law and international human rights law."

She pressed the UN to halt the "human rights violations," to protect minority neighborhoods from being "dumping places for industrial waste," and to force multinational corporations to better treat their neighbors.

Xavier's Dr. Wright told the UN gathering that the eighty-mile stretch from New Orleans to Baton Rouge produces one-fifth of the United States' petrochemicals.

"The air, ground, and water along the corridor were so full of carcinogens that it was once described as a massive human experiment," Dr. Wright explained, and in the process "transformed one of the poorest, slowest-growing sections of Louisiana into communities of brick houses and shopping centers, but not for the African-Americans who had been there long before the chemical plants came. The narrow corridor absorbs more toxic substances than do most entire states. Poor African-Americans living in small communities along the river bear the greatest burden of this pollution. Yet, they have benefited the least from its existence."

The UN commission invited the panel back for a second day, after hearing from Margie and her colleagues-in-arms.

When she returned to Louisiana, Margie wondered if she was any closer to her goal. The UN commission in Switzerland had heard her. Would Shell Chemical of Norco, Louisiana?

One weapon was the bucket testing, and once they got the apparatus going, a few residents in particular took to the work with vigor. Percy Hollins, a longtime Diamond resident who lived on Cathy Street, rushed outdoors with bucket in hand at the first whiff of odor.

In June 1999, Diamond's Bucket Brigade contingent went back to work, reenergized by Margie's appearance in the international forum. Taking the five-gallon buckets outside their homes on a Saturday morning, the residents set to work, with the help of a Sierra Club grant that paid for the batch of buckets. They walked to Washington Street, between the ball field and the Shell Chemical plant fence, and there Percy Hollins put his bucket to work, surrounded by nearly a dozen neighbors, some of whom held their noses because of the smell wafting in the air.

"I'm sensitive to odors," Margie said. "My throat's already sore."

Gaynel Johnson covered her nose and mouth as the odor walloped her.

Margie looked at her. "Go home."

"I am going. I got to leave. I am getting ready to throw up," Gaynel said. She scampered through a neighbor's yard to get to her house, hurried inside, and grabbed her inhaler to help her breathe.

The local press was now giving the matter more attention. The *Times-Picayune*'s pages the next day included a photograph of Gaynel Johnson pulling her shirt over her nose and mouth, which residents often did to avoid the smell.

Next, the residents trekked to the Bonnet Carre Spillway, about three hundred yards from the plant, and took another sample for comparison purposes. Denny Larson—who would later form Global Community Monitor, an environmental justice and human rights nonprofit—supervised the work.

The bucket testing would now go on full-time and become more systematic, with five Norco residents anointed as the official "sniffers" responsible for alerting the person with the bucket whenever the air seemed foul, so the tester could head outdoors and take a sample of the air.

That Saturday morning, Shell Chemical saw the ruckus across the way and publicly betrayed little worry about what the community was up to.

The Diamond residents will find nothing foul in the air, a Shell official said. Nothing at all.

Thirteen

ALLIES

As the Concerned Citizens of Norco recharged its mission that summer of 1999, a fresh face arrived. Louisiana native Anne Rolfes, a former Peace Corps volunteer and veteran of environmental struggles as far away as West Africa, had recently returned from the region and an all-out war against none other than Royal Dutch Shell.

In West Africa Rolfes's focus had been on Shell's treatment of land and people—and on the very public hanging of Ken Saro-Wiwa, a Nigerian playwright and noted environmental activist. Saro-Wiwa had led impassioned protests against the damage Big Oil created on the lands and fishing waters of the minority Ogoni tribe of Nigeria. He fought for just compensation for the people and their despoiled lands from Shell, which he blamed for the environmental degradation. Saro-Wiwa drew international attention by contrasting the riches going to petroleum companies and government coffers with the stark poverty of the people living on the lands that produced such wealth. He led marches of thousands of people to make the point, encouraging the protestors to carry green leaves to portray peace, and drawing the ire of a government not accustomed to being questioned, publicly or privately.

The Nigerian government soon accused Saro-Wiwa of being responsible for the deaths of four Ogoni chiefs in May 1994, and in short order, sentenced him to death. The world outside Nigeria viewed this as sham justice, retribution for Saro-Wiwa's challenge to the twin powers of government and oil. Nigeria is Africa's biggest petroleum producer, and oil accounts for some 90 percent of the country's export earnings. After his guilty verdict, but prior to execution of the sentence, Shell publicly condemned Saro-Wiwa and the protest tactics of his Movement for the Survival of the Ogoni People.

From the start Saro-Wiwa maintained he was framed because of his opposition to the military regime of General Sani Abacha and to the oil industry. "We confront these deadly enemies with the only weapon which they lack—truth," he told his followers from his Nigerian prison cell. "We would have to be ready to suffer arrest, detention, imprisonment, and death, as the only alternative to the struggle is extinction."

On November 10, 1995, his protest fell silent. He and eight cohorts were hanged.

The White House, European political leaders, and human rights organizations across the globe reacted with indignation, and two key prosecution witnesses later admitted they were threatened and bribed to give false evidence, intensifying the outrage at Nigeria's version of justice.

From afar, critics railed against Shell, saying it had a moral imperative to use its considerable might to halt the killing. The playwright's brother later told *New York Times* columnist Bob Herbert that he had pressed Shell to stop the hanging, and been told that something could be done, perhaps, if Saro-Wiwa called off his protest marches and issued a press release, on his group's official letterhead, telling everyone that Shell had caused no environmental damage in Nigeria.

Shell acknowledged that one of its representatives had met with the activist's brother, but dismissed his version of events. Amid fears of an international business backlash, the company told other journalists that it did not support the execution, but the government of Nigeria controlled its own courts, not Shell. Ultimately, there was no denying the bare truth: a noted activist had challenged the government's environmental record and targeted a giant petroleum company, and that activist was put to death by the regime, which was a close business partner of Shell Oil.

Rolfes ventured to the neighboring country of Benin, where many of the Ogoni had fled, several years after Saro-Wiwa's execution. She interviewed refugees from the Ogoni region and put their stories into a report with a title meant to strike a nerve with the big company—*Shell Shocked.*

The booklet was designed to resemble a passport, with "REFUGEE Federal Republic of Nigeria" printed across the top; below that was stamped a bright red frame with the words *"Shell Shocked."* Inside the booklet were stories from the refugee camp of 950—800 of them inhabitants from the Ogoni region of Nigeria—who had been uprooted to Benin. The sustenance per month per person, according to the report, was four tins of tomato sauce, one packet of sugar, one kilogram of fish, four kilograms of starch and rice, one-half kilogram of beans, one liter of oil, and one spoonful of salt.

That same year, Royal Dutch Shell proudly announced that its projected cost savings from a 1998 restructuring totaled $2.5 billion.

"Royal Dutch Shell's irresponsible oil production in Nigeria has caused these Ogoni to become refugees," the pamphlet said. "Shell's quest for oil—the drilling, the pumping, the oil spills, the gas flares—has destroyed a once fertile Ogoniland. The Ogoni responded to this violent destruction with peaceful protests made famous by Ken Saro-Wiwa and The Movement for the Survival of the Ogoni People." The report also alleged that collusion between Shell and the military resulted in the crackdown on the Ogoni, who were then forced to flee the country to stay alive.

"These refugees are different. Shell-shockingly different. The Ogoni are corporate refugees," said Rolfes's report. "Destruction of people's land for natural resources is nothing new. But the extent of the social and environmental devastation in this case—the fact that a corporation's activities have forced eight hundred people into a refugee camp by rendering their homeland unlivable—this is indeed a new and shocking turn of events. Even for a corporation with the power, shameful history, and ruthless reputation of Shell, this case begs a new question of corporate accountability."

At the time she coauthored *Shell Shocked*, Rolfes was working for a group based in Berkeley, California, called Project Underground, whose mission is to support communities affected by mining and oil operations. Rolfes had found a mission in

West Africa; but she always knew she would eventually come back home to Louisiana.

Now sharing space in her brother's small uptown New Orleans apartment while she hunted for a new quest, along with a job that could produce a steady paycheck, Rolfes turned out one afternoon for Bucket Brigade training in another community, Lake Charles, and there she met Wilma Subra. "I'm trying to find a way to do this work for a living," she told Wilma. Wilma delivered a dose of reality: "It can't be done."

Rolfes hadn't known a thing about the Bucket Brigade until that day, but she was quickly impressed by the hands-on way that community members tested the air. This was real grassroots work.

Rolfes also soon met Bucket Brigade mastermind Denny Larson. He knew instantly where she might find an avenue for her activism.

"Oh," Denny said to her, "you should go to this community in Norco."

There she went.

Anne Rolfes has a round, pretty face and the doe-eyed look of youth, with short-cropped light hair and a voice that carries the grainy texture of credibility. Behind those eyes is a practical strategist, a woman sharp enough to find an edge to provoke big industry, dogged enough to press multinational corporations for change, and not a bit fearful of stirring their ire. She carries with her the unshakable belief that big business, and particularly Big Oil, has too long ruled its environs with impunity, and that companies that pocket billions of dollars in profit should be called to task for the harm they've inflicted. Ignore her and she'll press on until you hear her voice.

Mislead her and she'll call you out publicly. Disregard the communities she aligns herself with and you'll find yourself the focus of a global public relations campaign.

Rolfes's first contact with the Diamond community was during a meeting at the house of chief bucket tester Percy Hollins. She listened quietly as residents detailed the next steps they would take to get Shell to move forward on the community's quest to relocate, and then another day she knocked on Margie Richard's door. Rolfes found herself inside a tiny trailer buzzing with activity, with nonstop phone calls and friends and family popping in and out. But Margie found time to welcome this new face to Concerned Citizens of Norco, who sustained herself with the aid of a Sierra Club grant, support from a California nonprofit, and part-time work at a homeless shelter.

Rolfes sensed instantly that Diamond's fight needed to take a sharper, more focused form. She encountered what she called a "muttering skepticism" from some in the neighborhood. People knew they were being exposed to pollutants in the air, and cursed big industry as the cause of this harm, but still felt powerless. It was the same resignation Margie and Gaynel had met in their first door-to-door trips drumming up support for the relocation plan.

Her mission became to turn the muttering skepticism into agitated action. To do that, she had to practically move into Diamond, a community with few e-mail accounts, cell phones, or fax machines, making personal interaction the only real way to be embraced.

It didn't take Rolfes long to become rooted, and she spread what she had learned about Shell in West Africa. By September

1999, Margie was coincidentally part of a contingent that spent ten days in Nigeria's Niger Delta region, where she and other activists visited communities affected by the operations of Shell, Mobil, and other multinational corporations. Margie's activism had taken a global route, and the lessons she learned in Africa girded her even more back home in Louisiana.

In turn, Rolfes set out to learn all she could about Shell in Norco.

Rolfes's first big public showdown came November 4, 1999, little more than two months after her first acquaintance with the Diamond community.

The EPA was holding a meeting at a downtown New Orleans hotel, and Shell was present for the release of a yearlong study into how Big Oil could better get along and improve communication with the residents of Norco, particularly those from Diamond. Rolfes envisioned the two entities—big government and big industry—patting each other on the back about how hard they had worked to improve conditions along the Chemical Corridor, and indeed a Shell-affiliated company delivered a slide presentation on the steps it was taking to improve relations with Diamond residents. Shell vowed to hire more workers from the community and adopt more youth programs for its children. As always, Shell seemed to have the stage.

On this day, however, the Concerned Citizens of Norco eyed its own piece of the stage, renting a half room just across from the official gathering in an orchestrated move to publicly challenge Shell. Inviting the gathered press to attend, the group focused on a report examining life in Diamond and pollution from the plant, stressing the wish to have a Shell-bankrolled move away from the smokestacks and flares. The report—compiled by the Sierra Club, the Concerned Citizens of Norco, Xavier University, the Earthjustice Legal Defense Fund, and another group affiliated with bucket pioneer Denny Larson—included pie charts listing toxic air emissions in St. Charles Parish and bar graphs detailing air releases of recognized carcinogens from Louisiana petroleum facilities. The report concluded that Shell was the worst offender of all.

The activists hoped the Diamond residents might finally snare a piece of the spotlight that had always seemed to shine elsewhere. They were, at last, successful.

The next day's River Parishes edition of the *Times-Picayune* carried a giant front-page picture of several Diamond residents and Anne Rolfes sitting at a table at the conference. Above their heads was a hand-scrawled message, the title of the report handed out that afternoon, and which was now fully displayed across a front-page photograph:

SHELL NORCO
TOXIC NEIGHBOR

Scanning the morning paper, Rolfes knew that the stealth tactic had indeed interrupted any goodwill session that had been planned, and Diamond residents had upstaged Shell at its own event.

Even Shell's own words snapped back against it. A Shell manager had proudly told the gathered press that Norco was a fine place to live, and that he'd have no hesitation raising his children there; the manager actually lived miles away, in Kenner, a detail that made its way into some of the coverage.

November 5, 1999, marked a new day in the battle for media attention. When Rolfes first ventured to Diamond, she heard the message loud and strikingly clear: "You can't get press on the issue." Of course, there would be a headline or local TV clip from time to time, but Diamond's community was almost always a bit player. Now Rolfes picked up the River Parishes edition of the big-city paper, and she knew the critics were wrong.

The coverage came near the anniversary of Ken Saro-Wiwa's death four years earlier. Rolfes took note of the date and its meaning, and felt the community was onto a new strategy that could finally make a difference.

Just as Diamond's fight hit the front page, Shell faced a new spate of troubling headlines about its operations. Soon, the federal EPA would cite eleven chemical plants and refineries in Louisiana and Texas for half of the accidental releases in a five-state region clustered with literally thousands of plants. Shell Chemical in Norco was among them. The EPA had launched the study after prodding from Subra, who believed the government had a mandate to tell fenceline residents what was in the air, and who personally drove EPA officials around in her silver blue Chevy Lumina to stress the urgency of the matter. Subra wheeled from Norco to Convent to Baton Rouge and into Texas, pointing out one community after another, until finally a government official turned to her. "Wilma, you can back off." She had made her point.

Not long before the activists' November surprise, the Environmental Defense Fund, a New York–based group, issued a damning report against Shell's Norco operations. The group studied 144 refineries nationwide, and it found Shell Norco to be in the bottom 20 percent in overall pollution performance, and to be one of just 12 refineries to score the lowest possible rating in four of five categories related to clean air. The company's Norco plant had the distinction of being the lone Louisiana facility to qualify as one of the worst refineries in the country, and the report cited its release of benzene and other cancer-causing chemicals. The *Times-Picayune* reported the findings.

As it had a decade earlier, Shell went on the offensive. It disputed the study's methodology, and it insisted that its emissions posed zero health risk to the community. "We're not the best. We're not the worst. We're sort of a middle-of-the-pack refinery," a Norco manager told the local newspaper.

Even as it was downplaying that report, Shell was dealing with a fresh fine from the Louisiana DEQ, this time for failing to monitor fugitive emissions a year earlier at the chemical plant. The fine topped $66,000, as the state said the company had failed to check more than four hundred valves capable of leaking toxic chemicals. Again, Shell portrayed this as a minor problem in the context of its vast operations.

Rolfes took notice of the rash of headlines, and she created the nonprofit Louisiana Bucket Brigade to help steer the Norco grassroots campaign. She picketed Shell alongside Diamond residents and helped the community collect air samples with the simple but effective buckets, gathering statistics to challenge Shell's complacent stance. In 1998, state records showed, Shell's twin chemical plant and nearby refinery had been the source of 50 percent of the toxic air emissions for the entire parish of St. Charles.

As the findings accumulated, the community gained strength. Rolfes crafted press releases, contacted the media, organized rallies, and kept close score of Shell's statements and actions. Armed with mountains of information, Rolfes hoped to force Shell to fulfill its vow to be a "good neighbor."

Community activism flies on the wits of the participants, not on big budgets. Before every press conference, Rolfes would turn to the community and say, "Can you come? Can you come? Can you come?" Describing the response, she said, "Most of it was, 'No, no, no, no.' You'd get five people and a few more would come over when they saw the cameras." A colleague witnessed her bustle and said to her, "Anne, every day you make stone soup."

Since Rolfes was on the ground in Louisiana, Denny Larson would now fly into town from San Francisco for important strategy sessions and major events. Diamond's other allies also remained engaged. Xavier's Deep South Center for Environmental Justice was following up on its sickness study with more reports detailing poor health in Diamond, Wilma Subra kept analyzing the air, lawyer Monique Harden continued providing legal counsel to residents, and the Sierra Club of Louisiana kept digging for facts to provide to the community's grassroots movement.

Soon, an important new voice arrived. Michael Lerner, the president of Commonweal, an influential California nonprofit focusing on health and environmental research issues, became attached to Diamond's cause from the moment he stepped into the community. As part of their environmental health focus, Lerner and colleagues had taken a "toxics tour" of Louisiana, guided by Harden, Rolfes, and Subra. When they reached Norco and stepped into Margie's trailer, the chemical smell was so bad Lerner felt sick, and small bubbles appeared on his glasses. As Margie told him the community's story, the booming loudspeaker from the plant next door at times overwhelmed the conversation. Later she led him to the spot where two neighbors died in 1973, and Lerner took in the vision of this small community cornered by a chemical plant on one side and a refinery on the other.

"Wait a minute," he told himself. "This just isn't right."

Back home in California, his thoughts kept returning to Norco, and soon so did he. On another visit, he went back to the spot where the teenager and elderly woman had been killed in the Shell explosion. There, standing alone, Lerner became buried in thought. "I could not rest until we did something," he said.

A loyal reader of the *Economist* magazine, he recalled how Royal Dutch Shell had placed two-page ads describing the company's commitment to the communities it served. Lerner knew that was fiction, as far as Diamond was concerned, and he dispatched an e-mail to the CEO of Royal Dutch Shell. "I've seen your ads in the *Economist* and I've just been in Norco," he told the company, "and you wouldn't possibly know what's going on there, because it's not what you would want."

Lerner heard that his e-mail ricocheted through the corporate halls of international Shell. As he attracted even more foundation-circle colleagues to trek to Norco with him, word came one day that an official from the Shell Foundation in London wanted to come to town to meet the group.

Lerner and colleagues huddled at the company's Norco headquarters with the Shell Foundation official from London and company representatives from U.S. offices in Louisiana and Texas. Subra, like Lerner a MacArthur Foundation fellow, took part in the meetings, providing a statistical framework to further make the case for relocation.

"These people are really suffering," Lerner told Shell. The company's representatives listened intently but gave no firm concessions, at least not yet, to the community's goal of a full relocation. Yet Lerner believed the presence of the Shell Foundation executive showed that the company was taking the issue far more seriously.

"Shell London was promoting this process of engagement," he said, "and Shell U.S. was not accustomed to it."

With some doors now cracking open, activists kept the pressure up.

In one year alone, Rolfes logged more than 130 visits to Washington, Cathy, Diamond, and East streets. At least twice a week, Rolfes was knocking on someone's door.

When Anne and Margie went door-to-door in Diamond to drum up support from residents for the relocation fight, Anne noticed that Margie kept a bundle of used clothes in the trunk to hand out. During one meeting with out-of-towners, Anne watched as Margie greeted all with a down-home touch. She passed around a bucket seeking donations to help a neighborhood boy buy a bike. Some of the visitors were taken aback by the personal pitch for money, but that was Margie's way.

At times Rolfes witnessed the skepticism Margie had been encountering for decades. Fighting an enormous company like Shell was a battle you would surely lose, the naysayer concluded, so why bother?

Only later did Rolfes hear of the heat Margie had taken during the long fight, the gossip flung in her direction. "She'd go to the grocery store and she'd hear it from the white community and from the black community," Rolfes recounted. From the black: "You'll never beat Shell. You're in this for yourself—Shell is going to pay you off." From the white: "You're that uppity Negro."

As Diamond's focus sharpened and people outside Norco became personally immersed in the dispute, local Shell workers began keeping a more vigilant eye on events. On occasion, Rolfes suddenly found herself trailed by a red Shell security truck that followed her from Margie's to Percy's to Gaynel's, like a taxi waiting to usher its passenger to the next stop. One afternoon, Rolfes walked over and knocked on the security truck's window. "I'm done. I'm going to have a sandwich in town. You want to come?" Point made. Shell said it wasn't trying to intimidate but was just keeping a close watch on goings-on outside its sensitive chemical plant, as any corporation would do.

The local police were a presence as well. If several cars gathered around Margie's trailer for a community meeting, like clockwork the police would pass by time and again. It wouldn't take much, just a couple of cars parked outside Margie's property, and soon a police car would cruise by. The residents peered out the window to watch it pass. A few laughed at the obvious attempt at intimidation; some were angry. Margie looked up, saw the cruiser pass by once again, and went back to talking.

When Shell did finally begin meeting with the Concerned Citizens of Norco, it was typically in the white part of town, and often with a beefy police contingent on hand. One night, in one such meeting, Margie turned to Shell's David Brignac, the recently appointed manager of sustainable development in Norco, and someone who seemed, finally, to be listening to the Diamond residents' concerns. "Can we meet in the black community, and can the police please go out there where some crime is being committed?" she asked Brignac.

Even with the small but encouraging steps forward in Diamond's seemingly never-ending quest for relocation, the truth was, the battle was draining Margie, and some of the consternation continued to come from her own neighborhood. As residents from Washington and Cathy marched in the streets and peppered Shell with questions, many residents from the back two streets, Diamond and East, remained skeptical that "crazy Margie," as some jokingly called her, could make Shell do a thing.

"Who do you think you are?" the skeptics asked. "Shell hasn't done this in years. We can't go against them."

One evening Shell was meeting with Diamond residents at the Norco Social and Pleasure Club, a meeting hall on Diamond Road that hosted student dances and repast dinners after funerals. Margie's blood pressure had gone up, and she was spending considerable time tending to her eighty-year-old mother. She was not planning to go.

"Listen, if God called you to the task, you stay focused," Mabel told her. "I'll be all right when you come back."

At the hall Margie pressed Shell again but heard little, if any, support from the people on Diamond and East. During a fifteen-minute break, with Shell officials stepping out of the hall, she knew it was time to confront her neighbors.

"A house divided against itself cannot stand." She challenged them. "I am upset with you all for not being with us in the beginning. We *all* have the same struggle. When Shell comes back, you have two things you can do. You can continue to bicker and get nowhere. Or you can come together and the victory is around the corner."

When Shell came back into the hall, there was unity at last, with people from the back two streets joining the chorus for relocation.

During trying moments like this, Margie gained inspiration from the mother of the boy burned to death in 1973, Ruth Jones. Mrs. Jones was aging and weak. Margie and Anne Rolfes wanted to hear from her directly about the explosion that, nearly thirty years later, resonated still in Diamond. Early one Sunday, Margie saw Mrs. Jones and asked, "Can I come by and talk to you about something?" After services at the redbrick Greater Good Hope Baptist Church on Cathy Street, where Ruth was a church steward and part of the mission band, Margie walked to her tidy brick apartment on Diamond Road.

"Well, Mrs. Ruth, how did you feel toward Shell?"

"Baby, it happened," Ruth told her. "You have to forgive. Honey, God's in control. You can't hate."

Moved by this woman's ability to forgive, Margie next turned to Shell, seeking the company's side of the story about that day in 1973. "How could a disaster like this happen—and I don't remember anyone talking about a recovery crisis program," she said. "How could you be so cruel?"

The Shell officials told Margie none of them were working in Norco at the time of the blast, but they steadfastly maintained that the company had acted properly in 1973. "How did you do well," Margie replied, "when nobody in the community knows what you did?"

Rolfes also visited Ruth Jones, who rarely ventured outside

her home. Ruth told her much the same story she had told Margie, and she did so without flinching. Rolfes learned that the woman had never filed suit back then. "That wasn't in a time when we knew about lawyers," the elderly woman informed her.

Mrs. Jones confided to Rolfes how Shell had come to the wake after Leroy's death and handed her $500. From this intimate conversation, the community learned, for the first time, after all these years, of the big company's small gesture. This fresh discovery of a historic outrage made its way into the activists' battle literature, one of many new weapons in their full-throated fight for change.

In the past, when Shell said foul emissions were rare, the community had little proof other than its collective recollection. Now Margie and her neighbors were creating journals logging each time they shut their windows to avoid the smell, and each time someone in their home turned ill.

"A noise like a whistle blew, then I looked out the side window or door and saw a lot of smoke. I got scared and called the Sheriff's Department. They said it was steam. It was so large I called DEQ to see if they had any other information. I then went to the front yard to see, then I was frightened because the steam was heavy + thick," Margie wrote one December morning in 1999 around nine. "I'm just sick of this place."

A month later, 2:30 in the morning: "The flame was so high and loud it caused me not to sleep." Three days later: "Loud noise like a roaring bomb. The noise from the flame is constant."

When Mabel Eugene's asthma worsened again, Margie made the connection. She looked up the federal Clean Air Act, and concluded that it was not being enforced in Diamond. Margie challenged the federal government, saying if the law truly were being applied, her mother and neighbors would not be so labored in their breathing. This, too, reenergized the Diamond residents to log all the foul smells, all the calls made to authorities or to Shell over the quality of the air and the sick people in their homes. Their home-scrawled dispatches showed every detail of their life next to the Norco plant.

In April 2000, another mishap prompted Gaynel Johnson to jot her own journal entry on the official letterhead of Concerned Citizens of Norco, whose address was 28 Washington Street, Margie's trailer home. "I shut all windows and doors," she wrote. "The odor got in the house and I couldn't breathe."

For Gaynel, the constant trips to the doctor were evidence enough. It often began with tightness in her chest, and soon enough, she was back at the hospital.

"Ms. Johnson, whenever those odors are outside, don't go out there," the doctor told her. Now living on Diamond Road, she couldn't sit on her porch anymore, and thought hard before working in her flower bed.

"A lot of times, I didn't have to go outside," she recalled. "They were old houses; it would seep into the house. I would start choking, gagging, and my children would say, 'Come on, we have to take you to the hospital.'"

Prompted by such daily aggravations, the residents kept pushing Shell. At one point, the community held a meeting with Shell in the house of chief bucket tester Percy Hollins. The company's team included an African-American official who reiterated how safe everything was, and in days past, some in Diamond may have taken Shell at its word. But with the bucket testing bringing its own truths, skepticism about Shell's message filled Percy Hollins's house.

By now, a new weapon was added to the community's arsenal. The "FlareCam" had come to Diamond, and it pointed directly toward the chemical plant's smokestacks and broadcast live on the Web for anyone to see.

The brainstorm had sprouted when the Sierra Club's board of directors came to Diamond for a tour. As they gathered in the playground across from the Shell chemical plant, Sierra Club leaders detected how bitter the fight had become, and one later wondered if it wouldn't make sense for the world to see for itself what was going on. With the FlareCam pointed right at the plant's flare stack and broadcast in real time, people outside Norco could come to their own conclusions.

Great idea, but where would the camera go? Who in Diamond had enough gumption to allow the contraption to be erected on his or her property and pointed across the way to Shell?

Margie Richard raised her hand. "I'm shaky about it, but it must be done."

Sierra Club senior regional representative Maura Wood installed the camcorder on Margie's house. "Her trailer was actually the closest to the flare that had a view. We wanted to put it someplace that was somewhat sheltered. By putting it under the trailer it kept it out of the rain," Wood explained. "She was willing to put in the extra phone line so we could get the Internet connection. And she was the person we were working with as the leader of the group at that point."

In 1999, the Web site of the Concerned Citizens of Norco began a one-year experiment in which viewers could click on and see a real-time image of the flare stack at the plant. "Flare-Cam: SHELL-Norco EXPOSED" was the site's headline, and activists spread the word among environmentalists about its existence.

"If the flare is burning large, yellow and smoky, the citizens on the fenceline of the plant are experiencing noise, smells, and the fear of breathing chemical pollution," Web viewers learned. "They live in fear of the next deadly event."

It wasn't long before a Shell security truck was passing back and forth in front of this new device. "You can look all you want—this is my yard, this is my property," Margie told the security officers.

The Sierra Club dug deeper into Shell's records and found that there had been an average of one accidental release a week. Maura Wood watched events unfold and was moved by the Diamond community's mission. "It was clear that the people had just reached a level where living next to that anymore wasn't tenable for them; they didn't feel safe. They were concerned about the day-to-day impacts and the major impacts in the event of a major accident."

On the first day of December 1999, nearly one full year since the scare at Norco Elementary had reinvigorated the community's mission, Margie Richard took to the podium at the fourteenth annual meeting of the EPA-created National Environmental Justice Advisory Council (NEJAC), this day gathering at the Crystal City Hilton Hotel in Arlington, Virginia.

Margie appeared with the North Baton Rouge Environmental Association, another group she had joined in her quest to unearth environmental truths about Norco and Shell. Other community activists rose to speak first. The Concerned Citizens Committee of Memphis, Tennessee, challenged the operation of a federal defense facility. The Citizens Against Toxic Exposure of Pensacola, Florida, complained that residents were not being relocated from contaminated Superfund sites in the city. The People of Color and Environmental Health Network cited pollution problems at former federal nuclear weapons production sites.

Finally, it was Margie's turn at the podium.

"I'm asking that you do whatever you can to have us properly relocated," she said. "We have suffered tremendously from death, respiratory diseases. We have suffered tremendously from being racially discriminated against and treated as less than human beings.

"Norco is divided by racism," she continued. "Nothing has improved. It has been proven that we are inhaling over twenty known carcinogens that come from the plant weekly. . . . So please, we're asking, we're begging, we need help. We've been to local officials, state officials, national officials, and internationally. We're asking that you would consider and let this be the ending of the pollution for the residents living in that area and may the year 2000, the next millennium, bring us some type of peace so that we can live just like everybody else wants to live."

As soon as her speech concluded, the council members had

questions for Margie, one wanting to know whether Diamond homesteaders had succeeded in securing better routes out in case of emergency.

"We asked for it," Margie told them. "We've come to the table negotiating for it. But let me say this. . . . It's very difficult. Norco has been known for having the best evacuation equipment there is in the world. But it does not meet the needs of the people in the Old Diamond Plantation. They have not come up with an alternate route."

Another member wanted to know if the community had enough training in how to deal with an emergency evacuation. Margie told them the training was inadequate, and that when disaster visited again, chaos would surely ensue for the elderly and infirm in a community with scant exit roads.

This time Margie's speech grabbed the attention of an official with Louisiana's Department of Environmental Quality, who publicly vowed to sit down with Diamond leaders to hear the community's story firsthand.

Even more, the Louisiana environmental agency would soon begin scrutinizing the operations of the Shell refinery at an unprecedented level. This time, someone from within the company would have the government's ear.

Fourteen

———◆———

THE
WHISTLE-
BLOWER

Barry McCormick was no phys-
ical giant, standing five feet four. A blunt-talking southerner
from Alabama, McCormick could be abrasive, even on good
days.

Yet behind the rough edges and rich drawl stood a highly
intelligent professional engineer who would unsettle many
around him at Shell's giant refinery.

McCormick was hired in 1997, around the time Shell was
beating back the lawsuit by Margie Richard and her neighbors. Even if their paths never literally crossed, it would have
been unimaginable that this company insider couldn't have
known or at least heard of how this woman from Diamond was
the thorn in Shell's side, pressing their chemical plant for decades on behalf of her neighbors and alleging a litany of environmental ailments. All of this would have been more than

relevant to McCormick, and if some of his colleagues were dismissing Margie's pleas, they wouldn't be amused for long, as McCormick himself began raising his own environmental flags.

McCormick earned $75,000 a year as Shell's senior environmental engineer and air program coordinator, and his mandate was to ensure that the company's Norco refinery met regulatory rules and clean-air regulations. McCormick did his job well, informing his bosses when they needed to reduce toxic emissions and take other steps to curb pollution, and dealing with state inspectors when they came to examine the plant. The longer he stayed, the longer his to-do list grew, and some of his coworkers and bosses were getting fed up with his constant reminders to curtail the leaks seeping from the refinery's countless valves, pumps, and seals. His bosses told him the plant was operating just fine, a point generally confirmed for years in the state DEQ inspection reports.

McCormick wasn't convinced, and one day he decided to make his point in dramatic fashion. Walking into a Shell conference room filled with colleagues and company brass, he slammed a metal bucket on the table. The bucket was used to catch chemicals that had been leaking from one wing of the facility, and McCormick pointed out that you wouldn't need a bucket if you didn't have leaks. Some of his coworkers felt he was showing them up and darted right out of the staff meeting in a fit of protest against him. The story made it into the *Times-Picayune*.

Soon enough, it dawned on Shell that this firebrand didn't fit the company profile, and Shell moved to extricate itself from him. But first, McCormick would later say in court papers, the company told him he could stay in his post while he looked for new employment, so long as he kept quiet about his belief in Shell's lack of compliance.

McCormick kept talking, informing the Louisiana Department of Environmental Quality about the dangerous emissions spewing from the refinery, renamed the Motiva Plant after a joint venture between Shell and Saudi Aramco.

On June 2, 1999, after learning of McCormick's discussions with the state DEQ, Shell banished him from its premises. Before the company escorted him out, one boss mentioned McCormick's focus on ethics. He just didn't "fit in" at Shell, was how McCormick remembered it, when he confronted the company not long after this escorted exit.

After Shell showed him the door, McCormick visited the downtown New Orleans law office of Jay Alan Ginsberg. Ginsberg specialized in employment labor matters, and his law practice stood a straight shot from the Superdome exit off I–10. The building was quaint, with old wood fixtures, and a framed New Orleans jazz poster hung on the wall in Ginsberg's office. On that first visit, accompanied by his wife, Barry McCormick was angry. He had done the job Shell had hired him to do, and done it well, and for his trouble his income was severed, his credibility was cast into question, and his life was rattled. McCormick displayed no second thoughts about confronting Shell.

He also spent even more time with the state DEQ, detailing how Motiva Enterprises was concealing the truth about the fugitive emissions at a plant so large it could now process 250,000 barrels of oil each day.

McCormick's words brought newfound attention to the refinery, and the state DEQ would unearth blatant pollution violations that for years had been overlooked. Digging deeper, regulators discovered some frightening near misses. In 1996, a large cloud of propane had been released in the area, dissipating before it ignited, averting yet another catastrophe. This was just the beginning of what the government would find.

In June 2000, as the DEQ looked closer at the refinery, McCormick made a move he hoped would both reestablish his credibility and make Shell pay for damaging his career. He filed suit, and there was no mistaking the seriousness of *Barry McCormick v. Shell Oil, Motiva Enterprises, et al.* His lawsuit asserted that he hit a brick wall from his boss when he tried to clean up Shell's environmental record, and that "in retaliation for his efforts," the company delivered a pink slip.

Though McCormick "was one of the most ethical people that he has ever met, he did not fit in at Shell," the lawsuit quoted one company executive as saying. Shell concluded that he had not met performance expectations, but McCormick portrayed this assessment as a ruse to justify firing him, observing that the company never disciplined him or gave him a negative evaluation. "In fact, prior to his whistleblower activities, McCormick received positive evaluations and a promotion," his lawsuit said. Put simply, Shell fired McCormick for doing his job.

Shell/Motiva initially talked tough, saying the lawsuit would not stand on its merits and portraying McCormick as a "disgruntled employee." Yet behind the scenes the company knew it had a battle on its hands. Shell was in trouble, and the lawyers quickly worked to settle the suit before even more damaging details could come to light.

That same month, June 2000, spurred largely by the issues

McCormick had brought to its attention, the state DEQ issued a press release detailing a string of alleged serious violations. Headlined "DEQ ISSUES NOTICE OF POTENTIAL PENALTY AND ADMINISTRATIVE ORDER TO MOTIVA NORCO," it documented Motiva's failure to adequately report emissions from its Norco complex and to properly tag and repair leaks in a timely manner, and described environmental lapses during a barge-loading operation. "These items and others contained in DEQ's notice to Motiva are very serious and we expect full cooperation from the facility on addressing these environmental concerns," the state said.

The company continued to take other hits. John M. Biers, an enterprising energy reporter then with the *Times-Picayune*, disclosed that for years Shell had said its refinery emitted few, if any, leaks, known in the industry as fugitive emissions. After the DEQ stepped up its enforcement, Shell/Motiva changed its tune considerably, going from reporting zero valve leaks in the first six months of 1998 at one of its units to more than 220 in the six-month period just one year later. "One thing is clear: McCormick has single-handedly changed the dynamic between regulator and regulated at the plant," Biers's piece concluded.

The refinery and nearby chemical plant encountered 561 accidental spills from 1994 to 1998, Biers's front-page report noted, yet those problems drew only scattered, tepid fines from the local DEQ. With the firestorm from McCormick's suit building in the press, and with Diamond activists garnering the EPA's support at the chemical plant, the government was taking its most serious look ever at Shell's operations— and the state was eyeing a far more severe punishment.

With Shell backtracking and finally on the wrong side of the news, Margie and her neighbors knew the complaints they had lodged for so long were gaining currency. This time their concerns had been validated by one of Shell's own, and finally by the government regulators who for so long had said all was fine.

———

Fifteen

THE
OFFER

Forced into an uncomfortable corner, Shell suddenly realized it was time for a change not only in tactics but in faces.

Don Baker, the well-coiffed community affairs representative long viewed as a nemesis by Margie and some of her neighbors, was replaced by Lily Galland, a veteran Shell official raised in the shadow of the refinery on Good Hope Street, on the white side of town, who seemed to know almost everyone in Norco. Galland had worked for Shell for more than two decades in community relations—fielding calls from Norco residents, putting out newsletters, and organizing United Way fund-raising and educational outreach with area schools. A woman with a relaxed smile and a soothing cadence to her voice, she will never forget the middle-of-the-night phone call in 1988 after the refinery exploded, and remembers how the company worked around the clock to stabilize the community after the tragedy.

"Norco was a quiet town and we all got along. And the plant was always there; I never saw it as a nuisance. I never experienced the things the Diamond residents shared," Galland explained. Her father never worked for Shell but many of her friends' dads did, and she saw how committed their families were to the company. Yet in Diamond, if no one worked for the plant, there were no ties.

"I could see how neighbors felt: 'We don't agree Shell is a good neighbor,'" Galland said.

Galland tried putting herself in the shoes of black Norco, setting appointments to talk face-to-face with Margie Richard and her fellow activists, and the thought hit her instantly: Shell should have been reaching out to the Diamond community much sooner. "It was an education for me to be at the table to hear how some of the residents felt. A lot of our employees couldn't understand what was going on, but they work here and they have a tie to the facility and they don't fear working here," Galland recalled. "I feel that Diamond was ignored. I'm not just saying Shell ignored them, but they were ignored by other entities, including government. . . . Diamond had some real social issues and these issues weren't being addressed. The thing of it was, nobody was addressing their needs."

Attempting to mediate the decades-long discord between Shell and Diamond was the most stressful time Galland had experienced as a Shell employee. It wasn't just Shell, she reminded herself, that had not reached out to its neighbors, but an antiquated corporate philosophy that argued businesses shouldn't engage communities that eye them with suspicion.

Working closely with Galland were two key Shell figures. One was veteran Norco plant manager David Brignac, who in 2000 was appointed as manager of sustainable development. The other was Wayne Pearce, the influential site manager for Shell Chemicals Norco, who was brought in from one of the company's foreign offices, in the UK, also in 2000.

Brignac, with his dark hair cropped neatly and his face exuding a youthful openness, was approachable to residents. "Even though we disagreed about a lot of things, we were talking and he wasn't ignoring us," Margie observed. "He did not shun people. He did not close the door."

As the hands-on site manager, Pearce was more important than both Galland and Brignac, as his word carried sway with headquarters in Houston and London. Residents detected a Scottish brogue in Pearce's voice, and they came to know him as a reasonable figure who shared personal anecdotes about family vacations and never came off as arrogant. Still, it was clear this youngish executive was no pushover, and his interests were firmly with Shell. "It was him that had to say, 'Yes, we are going to do this or no we are not,'" said Wilma Subra, who was trying to bring final closure to the dispute. "And it was also very clear that he had to get the agreement of corporate."

Shell desperately wanted an end to the nasty headlines about its Norco operations, and it pressed the trio of veterans to find a solution that would make the headaches go away without compromising the company's standing or admitting that the company was to blame for the sick neighbors residing in Diamond.

Yet even with Galland, Brignac, and Pearce stepping beyond the razor wire, people in Diamond refused simply to take Shell's word that it would improve living conditions in Norco.

During a meeting one afternoon, Gaynel Johnson heard from Shell, again, how the chemical plant was not making people sick. Just look at our workforce, Shell said: healthy, happy, and faithful. These claims brought Gaynel's blunt demeanor to the surface. "We'll give you our key for one day and you give us your key for one day, and see how much you like living there," she told Shell officials. "Working there and living there are two different things. When you work there you have these safety outfits, these safety masks. We're not protected."

Was it too late for relations between Shell and Diamond to heal? Could a new corporate outlook change what the thirty previous years had wrought?

In the early fall of 2000, as Shell worked toward settling the high-profile whistle-blower case, the company formally issued an invitation to buy Diamond homes nearest its chemical plant. Shell was offering more money than it had heretofore been paying, and its proposed buyout also included homes on Good Hope and Norco, two streets on the white side of town.

Considering the company's long history, this sudden reversal of tactics proved both puzzling and bittersweet. But this new turn arose in the context of the legal battles Shell was facing. By going to state regulators first, McCormick had ensured fresh scrutiny of the company's Norco operations, not to mention potentially huge fines against Shell and its affiliates and a just payday for himself.

Faced with an avalanche of legal troubles, Shell may have offered to buy Diamond homes as part public relations strategy. The oil conglomerate repeated a vow to the community that it would reduce emissions at the chemical plant by 30 percent in the next three years, and reduce flaring by 50 percent. Yet these concessions had virtually been guaranteed by the pending DEQ settlement, to be released months after the whistle-blower's complaints. Shell's hand was being forced by a more vigilant government, which had just now confirmed what residents had been saying for years.

Either way, Shell was under attack when it offered to appease the Diamond residents. Yet what the community hoped would be a complete victory was only halfway there. Shell's plan, after all of the legalities had been settled behind closed doors, included the buyout of the two streets nearest the plant, Washington and Cathy, not the two farthest away, Diamond and East.

As Diamond neighbors were often blood relatives, Shell's proposal felt like an offer to tear a family in half.

"We have made it clear to you that we want the entire Diamond Community, which includes Washington, Cathy, Diamond, and East streets, relocated out of harm's way," Margie wrote Shell Chemical's plant manager on September 1, 2000. "We are all exposed to the toxic pollution leaked, spilled and flared from the Shell Chemical factory. On all four streets, we suffer from the same health problems, smell the same bad odors and we could all die the same death if there was an explosion at the Shell factory. We want a community relocation that is just and fair to everyone that lives in the Diamond Community. This offer is essentially the same you have been making since 1973."

That long-standing practice had been striking for its simplicity. The company would pay exactly what the property was worth, not a cent more, leaving sellers with little cash to look for a new home in neighborhoods removed from big industry. Most refused to sell under this entrenched company plan.

The company was now offering more money, but only for part of the neighborhood. Margie and the residents most active in the fight with Shell wanted more.

Diamond residents needed only to look at their kin to see what the plant was doing. While none of the individual chemicals released from the plant exceeded state pollution standards, Margie noted that multiple toxins had consistently cascaded into her neighborhood, and she felt that the cumulative effect posed far greater health risks than Shell acknowledged; chemist Wilma Subra agreed.

After Shell had laid out its offer, Anne Rolfes released a report that rivaled *Shell Shocked,* her dispatch from Nigeria.

For some time now Rolfes had been going to the courthouse to pull property deeds that showed how much Shell had paid for nearly fifty Diamond properties over the years. She interviewed Diamond residents about their own personal histories and views of Shell, then went back to her apartment in New Orleans to incorporate the information into a booklet, which she created with the help of Inno Nagara, a graphic designer she knew in Berkeley, California.

Shell Games: Divide and Conquer in Norco's Diamond Community; the Case for Fair and Just Relocation was issued in the late summer of 2000, and it began with a brief history of the four-street Diamond community, sitting in the heart of Cancer Alley and once the site of the Diamond Plantation.

"The people of Diamond can tell you where they were

born, where their parents were born, and where their grand-parents used to live in days gone by. Belltown, the Big Store, the Big Yard—these are the cherished names of places that are now occupied by the Shell Chemical facility. Sadly, since Shell entered the community in the 1950s, many of the places rich in the history of this African American community have been bought by the corporation for its expanding industry needs."

In blunt language, the report said Shell's proposal "appears to show how little the corporation understands its neighbors. Many families have close relatives that live on all four streets."

It cited the Hollins family, from Cathy Street, as a case study. Mary Hollins is the principal caretaker of her elderly mother, who lives on Diamond. "If Mrs. Hollins were to fol-low Shell's plan, she would leave the neighborhood and aban-don her mother. Does Shell really expect that Mrs. Hollins and others in similar situations will sell out their families?" the booklet asked.

The report cited the offer's timing, arriving as the EPA and DEQ were deep in their investigation of McCormick's story. "Is Shell using Diamond in a public relations ploy," it asked, "or to make peace with federal regulators?"

And then the report included a two-page diagram of every single property in Diamond, all 269 lots, and the names of the owners for each.

As Rolfes prepared the report, Nagara saw all the property information she had pulled together, and a brainstorm hit him. "Anne, you've got to put all the names on that map. You have to put the people's names on the plots, because it will give people a greater sense of community," he said. Rolfes knew he was right.

Of special interest was lot 22, square 14 on Washington Street. "On a day like any other in 1973, gas shot from the Shell facility and into her home. Mrs. Washington was killed instantly. Leroy Jones was cutting the grass in her front yard. His body was set ablaze. He was taken to the hospital and died days later. Lot 22, Square 14 is today a sacred place, a sad place, a testament to the urgency of the Diamond community's demands."

Shell Games reported how on February 3, 1977, Mrs. Washington's sister sold the empty lot to the company for $3,000, part of a long-standing practice of properties selling for bottom dollar. Over the years, Rolfes found, Shell had been buying empty lots and mobile homes for an average of $10,000, wood homes for about $20,000, brick homes for $40,000—an average overall price of just over $24,000 per lot.

On the final page, the activists displayed a colorful picture of a pretty house in the suburbs, its two stories lovingly shaded by trees. "Unlike the people of Diamond, the manager of the Shell refinery lives in a neighborhood far from large industry and pollution. He has chosen to live in clean air, about twenty-five miles away from the Diamond neighborhood and Shell facilities."

The map, the text, and the images of *Shell Games,* shared with residents and made available to the press, government, and other activists intrigued by Diamond's battle, proved to be powerful on multiple fronts. One, they displayed the breadth of the research the activists had conducted. Two, in compelling language, Rolfes laid down a gauntlet to Shell: We're not taking your word anymore. And three, by putting the names of every property owner onto a detailed map, the booklet brought together all four streets of Diamond, the family tree now spread out for all to take in. The company surely had to notice what Anne Rolfes, Margie, and her neighbors were saying, and it soon reworked its offer.

Four days after Margie's September 1 letter, Shell announced an "enhanced Norco Voluntary Fenceline Property Purchase Program" that now offered "Market Value plus a premium," along with provisions for moving expenses, rent disruption costs, lot clearance bonus payments, and guaranteed minimums for properties.

"The Norco Fenceline Voluntary Property Purchase Program is an enhancement of a thirty-year effort by Shell Chemical Company and Motiva Enterprises LLC to purchase residential property along the East and West fencelines of their facilities in Norco," Shell said in the official "Revised Program Book."

Shell said the program was intended to accelerate the purchase of remaining properties to develop "a 'green space' that can be used by all Norco residents. The vision for the greenbelt includes recreational parks, walking trails, and structures to improve the beauty of the community."

On Washington and Cathy just off the West Site's chemical plant, Shell offered a minimum $50,000 for residential properties, and half that for vacant ones, plus bonuses including moving expenses up to $5,000 and a legal consulting allowance up to $500. Near the refinery on the East Site, Shell made a similar offer along the east side of Good Hope Street—

which stood closest to that Shell facility—and single-family residences on one side of Norco Avenue.

After three decades, the company had finally, and formally, ditched its firm stance of paying "fair market value" in Diamond. Its "Revised Program Book" included worksheets showing residents how the numbers had changed. A property appraised at $25,000 would for years have sold for precisely that; now, with premiums added, the guaranteed floor of $50,000, and potential lot clearing, moving, and other bonuses, it could go as high as $60,500. A property appraised at $50,000 could yield $75,500, and so forth.

Shell said its offer was consistent to all property owners in the area, but as a voluntary program, the final decision was with individual homeowners. Diamond residents were given an extended deadline, until December 31, 2001, fifteen months, to make up their minds. Brignac said Shell would set up an informational meeting for the community to learn more details.

But even without offering to buy out all four Diamond streets, Galland, Brignac, and Pearce wondered, would the enhanced offer extinguish the fire?

They found out quickly. The answer was, resoundingly, no. While individual owners were free to sell to Shell, and some considered doing just that, the Concerned Citizens of Norco and their allies quickly let the company know this was not the solution they had sought.

The Diamond community realized the offer was a step forward after decades of setbacks; but the residents also knew they were onto something. The grassroots campaign, now more energized and formalized, had finally moved the giant.

Why stop now? Anne Rolfes, Margie Richard, Monique Harden, and colleagues wouldn't give up until all four streets were included, and once again they pressured Shell. By the summer of 2000, the company had announced the Good Neighbor Initiative, which it promised would be backed by big investments in Diamond. Margie got a copy of the report and read it word for word, and then pressed Shell to make good on every single promise to reduce emissions and flaring.

That fall, during a presidential election season filled with debates, Diamond residents challenged Shell to a debate inside the Masonic lodge on Washington Street. The company declined to step across the street to take part. The debate went on nonetheless, with residents hoisting pullout quotes made previously by Shell brass contending the company's offer was fair and the community's health was fine. A dozen residents challenged these Shell statements with stories of how Diamond's history had been tainted by sick children, dwindling property values, and an incomplete buyout. Margie argued that the offer should be enhanced to compensate residents for the sickness in their homes and for the history that was bulldozed when Shell built in the first place.

"So far they are not negotiating, just dictating," Rolfes told a local reporter. "They keep saying the community doesn't understand our proposition, and here they had the chance to explain. They only go to the meetings that they control."

Three days after the one-sided debate, the *St. Charles Herald-Guide* ran a headline that surely made Shell wince: "Shell Ducks Debate with Diamond Community." A sidebar article that began "What would a presidential debate be like if one of the candidates didn't show up?" ran alongside the report. It was another illustration, Rolfes knew, of manufacturing momentum.

In November 2000, as the offer to buy only half of the Diamond neighborhood persisted, Royal Dutch Shell reported record profits of $3.25 billion for the three months ending September 30. "The result was above a range of analysts' expectations and the biggest profit Shell has ever turned in," worldwide press reports said. The Diamond residents wondered why Shell, with its bank account flush, couldn't pull out the equivalent of spare change and extend its offer to all four streets. It was the same question the community's lawyers had asked a decade earlier.

That same month, the activists launched a new international offensive against Shell, with a personal touch in mind. They decided to fly Margie to the UN Conference on Climate Change at The Hague, where Shell executives were making a presentation. Margie would arrive with a bag of air from Norco. She would present this package as a gift: a very public challenge to a company so concerned about its image. "I want to put it in the hands of the person who is one of the owners of the Royal Dutch company," Margie told her neighbors.

On November 17, 2000, Margie gathered the bucket and her baggage to begin this long journey. Stepping out of her trailer to head to the New Orleans airport, she was greeted by a loud, fiery flare from across the street. As Continental Flight 388 took to the air, she looked out her window at row after row of billowing industry clouds below.

Margie carried the precious cargo all the way to Holland.

Finally she touched down, then made her way to the conference, the bucket closely tucked to her side, passing it through an X-ray machine and past skeptical security personnel at The Hague.

In the conference's main hall, she stood in the back amid an overflow crowd, as sharply attired foreign business executives sat around tables listening to presentations on climate change. When a Royal Dutch Shell executive finished his speech, Margie raised her hand.

Called upon, she approached the slim Shell executive with the bucket in hand. Some of those sitting nearby exuded disbelief as Margie, wearing a black dress and her hair tightly wrapped in a bun, pressed forth.

"I just wanted to ask if you're going to be true to what you say on paper. Are you going to be concerned about the lives of people? Now I know this looks strange but this is the bag that has been approved by government. This is a sample of the air that's been trapped."

Margie walked closer to the Shell representative.

"I came all the way here for this and it's really important. So I'd like to give you a gift of the air," Margie said.

"Can I breathe it?" he asked.

"I wouldn't take the chance," she replied, as Denny Larson snapped a picture of the exchange. "We're breathing it every day."

Afterward, in the hallway, Margie buttonholed the Shell official to tell him more about her hometown's struggle and to press him to "hear our cry." As he peered at her name tag, a light went off: This was Margie Richard. The Shell executive recognized the name.

"We'll be in touch," he told her.

Two weeks later, there was a knock at the back door of Margie's trailer. It was a representative from Shell's international office wanting to know more about the air she had flown to Holland to talk about.

As spring break 2001 arrived, Tulane University students rallied on campus on behalf of Diamond, with Margie in attendance as inspiration. The undergrads charged that Shell was putting profits before people, and they hoisted signs with Shell's logo and the words "Shame!," "$hell," and "FREE DIAMOND."

As summer approached, even more outrage arrived in the form of forty environmental leaders from across Louisiana and the country who marched through the streets of New Orleans en route to Shell's headquarters with cries of "Shame on Shell" and "Don't Divide Diamond." Rolfes led the charge, bullhorn in hand, walking beside activist Lois Gibbs, renowned for her fight involving the infamous Love Canal contamination case in New York in the late 1970s. "Shell, watch out, because the people are coming after you!" Gibbs shouted.

Twenty-three years earlier, Gibbs had learned that her neighborhood in Love Canal, New York, was located on a twenty-thousand-ton chemical waste dump, launching her on a crusade to force public officials to relocate the residents. The state of New York later shut down the local elementary school her children attended and purchased the 239 homes nearest the dump. Yet Gibbs believed those steps were not enough, and she kept the pressure on. In 1980, President Jimmy Carter issued an emergency order moving nine hundred additional families out. Now, two decades after that historic environmental struggle, Gibbs pressed for the same result for residents of Diamond.

Diamond activists were back on their computers, pulling together their latest missive—a four-page pamphlet titled *Shell's Great Divide: Lofty Principles vs. Low-Down Land Deals.* It detailed how in the first half of 2001, Shell personnel called the local Emergency Operations Commission to report releases, incidents, or accidents fifty-four times, or twice a week. *Shell's Great Divide* also stated that in 1999, benzene, a known and documented carcinogen, had been emitted into the air of Norco from Shell's plants, along with toluene and propylene. The report focused again on Ruth Jones, mother of Leroy, who lived on Diamond Road, a street not included in the company's buyout offer. "Despite earning a record-setting $3.9 billion in the first three months of 2001, Shell has excluded Mrs. Jones from the program," it said.

Shell now no longer ignored such provocations and held an informational meeting on the latest buyout offer at Norco Elementary School on Thursday, February 8, 2001.

Rolfes balked at the idea of once again meeting Shell outside of Diamond. Why couldn't the company meet in the neighborhood? Sure, Norco Elementary was nearby but the symbolism was in Diamond, and she suggested that the community not show up at all. Margie convinced Rolfes they had to go.

When Anne and Margie drove up together, there were police everywhere, including two intimidating officers, arms folded across barrel chests, at the front door. Margie recognized one from her middle school teaching days. "Hey, weren't you in my school class?"

The tension was broken, but the meeting failed to produce a tenable resolution. Afterward Rolfes wrote to Shell, "furious" that a strong police presence had been employed. Stop sending the cops to meetings with the community, stop patrolling around Margie's house, and pull back on the intimidation, she said. Shell got the message, and Brignac decreed there'd be no more cops, and soon the meetings would be held inside Diamond, at the Masonic hall.

Not long after Shell agreed to a meeting at the Masonic hall, the DEQ announced the results of its investigation into Barry McCormick's claims. The inquiry eventually led to the largest-ever settlement for the state environmental agency, focusing on an array of air quality violations at the Motiva plant in Norco and another Shell Louisiana facility in Convent. The Louisiana inquiry also became wrapped into a larger environmental case that combined the efforts of the EPA and states including Delaware, Texas, and Washington.

The Louisiana piece of the settlement was for "violating air and water quality provisions of the Louisiana Environmental Quality Act," the state said in a March 21, 2001, statement. An EPA announcement the same day called the settlements "part of an effort to reduce harmful air pollution released illegally from petroleum refineries." John Ashcroft, then the attorney general, hailed the settlements as a victory for the environment.

In its lengthy consent decree with the government, Motiva

Enterprises went on record denying the allegations against it. Yet "in the interest of settlement," the company agreed to install air pollution control equipment and enhance air quality practices at its four refineries.

In Louisiana, Motiva Enterprises was ordered to pay a $500,000 cash fine to the state, to fund a $280,000 cancer study, and to earmark $750,000 to develop an ambient air-monitoring network for three years and $3 million for a three-year flaring-reduction program.

"A significant number of the issues involved in the state of Louisiana's case against Motiva were identified at the Norco refinery as a result of information provided by a private citizen who was a former employee at the facility," the DEQ press release said. The employee, unnamed in the state's announcement, was McCormick.

A twenty-three-page settlement between the DEQ and Motiva Enterprises documented dozens of violations of air-quality regulations and state procedures, along with improper toxic releases, from the vast refinery. The two-year investigation zeroed in on two principles Diamond activists had long preached: Shell shouldn't pollute the air, and it should be truthful about when it does.

"The wide-ranging Norco probe centered on the company's failure to monitor, check and fix thousands of toxic leaks and on whether the company misrepresented its operations to the agency," the *Times-Picayune* noted.

The joint national settlement with Motiva called for another $2.3 million in cash and $1.25 million in environmental projects at plants in Texas and Delaware, and Motiva had to install state-of-the-art pollution-control equipment at all of its facilities. As part of a long-term plan, Motiva also agreed to spend $80 million to upgrade refineries in Norco and Convent.

Two months later, Shell formally settled McCormick's lawsuit with a confidentiality agreement that kept the dollar amount secret, ensuring that any further public relations damage was kept to a minimum.

Sixteen

"TOXIC PARK"

On a Saturday afternoon in May 2001, Diamond residents and activists gathered in the Diamond Community Hall for an event headlined as a "Pre-Mother's Day Lunch for All!" The invitations were adorned in bright colors, luring neighbors to a gathering where "all mothers in attendance will be honored and recognized. We are struggling for our children."

Honoring mothers wasn't the sole reason for this get-together. In truth, this gathering was as much about discussing the current phase of the battle against Shell, and the activists used the back side of the flyer to let everyone in Diamond know just how far they had come since the day the courts had rejected their lawsuit in 1997.

"Need Proof?" the paper asked. Then, to answer the rhetorical question, it outlined two columns headed "The way it used to be" and "The way it is now."

Used to be: Shell wouldn't meet with the Diamond residents or return their phone calls, declared once that "relocation is not an option," and often had a large police contingent alongside it during community meetings.

Is now: Shell speaks openly about relocation, changed the faces across the negotiating table, admits that its latest buyout offer needs reworking, and stopped asking the police to come to gatherings.

"The squeaky wheel gets the grease. If you live on Diamond or East, come to the meetings, make noise, <u>and call the bucket samplers when you smell an odor!!!!</u>" the flyer concluded.

In the summer of 2001, an added push came from Washington, D.C. Damu Smith, a leader with Greenpeace, arranged a celebrity bus tour of southern Louisiana's Cancer Alley, and among those in attendance were author Alice Walker, *M*A*S*H* star Mike Farrell, and California congresswoman Maxine Waters.

Margie helped host the out-of-towners. As children hoisted signs demanding clean air, Margie walked upon the grounds where Bethune High School had stood, inspired by the message the school's namesake had delivered years earlier: "Never start out to lose the race."

Congresswoman Waters was so moved by the visit that she called Shell Oil Company's CEO in Houston, Steven L. Miller, urging him to use his considerable might to get the company to buy out all four blocks. Encouraged this would happen, Waters wrote to Margie, copying Shell's Miller and Greenpeace's Smith.

"I consider one of the most important actions I have taken to be my visit to your community during the recent tour of Cancer Alley," the letter said. "I was saddened by the unfortunate situation confronting the people of all of the areas that I visited. I assure you, however, that I am committed to the enforcement of the Civil Rights Laws, in addition to other laws, in the environmental context."

The congresswoman shared Margie's ongoing optimism, telling her that Miller assured her he would rethink the company's strategy and seriously consider a four-block buyout.

It didn't take long for Miller to burst this bubble. While contending that the company was committed to making Norco a "thriving and dynamic community" and to "dealing fairly" with the residents, he told Waters point-blank: "The greenbelt we envision does not require additional buyouts."

Waters believed that the words she now read on official Shell letterhead were starkly different from the ones she had heard on the phone, and she wrote Miller once more, urging Shell to reconsider its "poor decision."

"I am disappointed in the apparent change in tone and direction of your letter from the telephone conversations we had had previously," she wrote, calling the company's stance "a clear example of Shell's insensitivity in ignoring the very real health threats that its chemical facility creates for African American people throughout the Diamond community."

California had always been a green state, fighting for everything from the rivers to the redwoods. Louisiana was all about black gold, where Shell's word was final, and a probing out-of-town politician would not change that.

Then, suddenly, Shell was backed into a corner of its own

making, piquing the community's anger to new heights. While the company was declining to buy out Diamond and East, it had just purchased property on the *other* side of the two streets, the grove of trees separating black Norco from white, in an effort to create more open space around its facilities.

The Diamond community discovered this hopscotched land deal by chance. A couple Diamond residents happened to be at an open meeting where Shell mentioned its purchase of the Gaspard Mule Line, the official name for the block-thick area of dense shrub and trees that separated the neighborhoods. It was a move that "really radicalized people," Rolfes said. "A final gut punch."

In August, the Bucket Brigade called another, particularly spirited, press conference on a plot of Diamond land. Homeowners, wearing shorts, T-shirts, and other casual attire on the summer afternoon, hoisted signs saying "We Shall Overcome" and expressed rage over the company's tactics. "You mean to tell me that grass and trees come before human beings?" Gaynel Johnson demanded.

Shell insisted it might well build a park at the Gaspard Mule Line, an attempt to offer an olive branch to Diamond. "You think we want a toxic park?" the residents wanted to know.

Toward the end of 2001, Subra released a report finding that during two weeks in June, Shell's chemical plant belched twice as much sulfur dioxide, benzene, and nitrogen oxide as allowed by the Department of Environmental Quality. Benzene, a known carcinogen, can cause convulsions, coma, and damage to blood cells in high doses. Sulfur dioxide can trigger a loss of smell and nausea, and coming in contact with it can burn the skin and damage the eyes.

Shell blamed the releases on power outages triggered by a heavy storm, which the company said forced the plant to release thousands of pounds of chemicals through flares, that familiar smoky visitor. "We had no choice," a Shell spokesperson told the press. "We had to flare." Again the company stressed that the flare did not harm residents' health.

Despite the company line, the probes into industry's questionable Norco practices continued to mount. In September 2001, *Environmental Health Perspectives,* a monthly journal of peer-reviewed research, published a study of the impact of petrochemical facilities on communities, like southern Louisiana. The results were precisely what Diamond residents had been saying all of these decades. Researchers found that in 1997—the same year *Richards v. Shell* was dismissed as a legal case by a St. Charles Parish jury—Louisiana ranked second in the United States in total releases of chemicals on the Toxic Resource Inventory. Cardiovascular disease and lung cancer were among the leading causes of death in Louisiana, and "the death rate from chronic obstructive pulmonary disease is 58% higher in Louisiana than the national average," the researchers found.

Even more troubling news for Shell came from Delaware that year. A tank explosion at a Motiva Enterprises refinery in Dover killed one worker, injured eight others, and spilled more than a million gallons of sulfuric acid, some flowing directly into the Delaware River, the Associated Press reported. The dead worker's body was not found. Investigators later discovered that the tank in question had a history of leaks, but Motiva had ignored warnings. Motiva later pleaded guilty to negligently endangering workers at the Delaware refinery and was fined $10 million.

As 2001 came to a close, more proof of the dangers Diamond residents had been complaining about abounded. There was no way the community would settle for an offer it believed was only halfway home.

Rolfes took to the computer once more, releasing *Family Divided by Shell,* whose cover featured a family photograph torn in half. It told how nearly 60 percent of families that were excluded from the two-street buyout offer "have family members on the other two streets." The following pages were devoted to profiles of family members excluded from Shell's offer, with stories of sarcoidosis and cancers.

Building from its discovery of Shell's plan to purchase the Gaspard Mule Line, there was a page devoted to a hand-drawn map marking the grove of shrubbery. "This is the Gaspard Mule Line, a grove of trees that has always separated the white and the black sides of town. Shell told us they couldn't buy Diamond and East Streets because they did not need or want any more property," *Family Divided by Shell* stated. "Why did they jump over two streets and buy the Gaspard Line? You mean Shell cares more about grass and trees than human beings?"

Still keeping an eye on Shell's bottom line, the pamphlet highlighted the company's latest profitable turn. "Royal Dutch Shell's earnings for 2001: $13.11 billion," it said, noting that a new Web site had been created—www.shelldividesfamilies.org—where even more information was available for anyone with access to a computer.

This pamphlet also was distributed to the community, to the press, and to activists in other fenceline hot spots.

Shell responded with a letter to the editor of the *St. Charles Herald-Guide*, saying that it was proud of its seventy-five-year history with Norco, that its plant emissions remained below air ambient standards, and that it was "committed to reducing toxic release inventory emissions by thirty percent." It said a greenbelt was being created as part of its corporate Good Neighbor Initiative. "This voluntary property purchase program is in no way tied to any health threats associated with any of our facilities," the company closed.

If Shell was such a "good neighbor," the people of Diamond wanted to know, why was it ignoring the pleas of half the community? Soon, fifty-four residents from Diamond and East streets signed their names to a petition demanding action, sending the missive directly to Shell CEO Miller on the community's letterhead: *CONCERNED CITIZENS OF NORCO. Prayer changes things.*

Ten days later, Miller heard from twelve members of Congress, led by Maxine Waters and including politicians from Florida to Michigan. "As you are aware," they wrote the CEO, "residents routinely have to rush their children and grandchildren to hospital emergency rooms because they are suffering from asthma attacks, some of which have required hospitalizations. Both old and young people have had to rely on oxygen tanks and prescription medicines to cope with Shell's toxic fumes. These people live on all four streets of the Diamond community, not just the two streets included in Shell's current buyout plan."

By this time, Harden and Rolfes had reached out to socially responsible investors in New York to inform them of Shell's environmental record. Hit the company in the pocket-book, they thought, and perhaps it will listen.

Harden had also done some digging into Shell's operations abroad, and one day she ran across a report from Shell itself in the Netherlands expressing "concern" that residents near one refinery were just one-third of a mile away. "They felt it was too risky and a hazard to have Dutch people one-third of a mile away," Harden explained. "I thought: One-third of a mile versus across the street? We've got a terrible double standard."

Dr. Beverly Wright, from Xavier University's Deep South Center for Environmental Justice, the group that had published a sickness study of Diamond in 1997, had continually been questioning Shell. Wright had been part of the contingent that traveled to Switzerland two years earlier telling the United Nations about "environmental racism" in southern Louisiana. Now, in a meeting with Shell, Wright accused the company of acting like "the Taliban."

Lily Galland cringed as she heard the words. Being dubbed the Taliban was an extreme sign of how far the sides had been apart all these years. "I don't feel the community felt that way, but I feel these environmentalists or activists, that's how they do business," Galland said. Yet the bare truth was: Had Shell reached out decades, or even just years, earlier, the gulf would likely not have been so wide.

Shell realized the pressure would not stop coming from the tight band of homeowners and activists. With it still rising, the company took an unprecedented step and signed papers saying it would sit down and negotiate, "in good faith," with the Diamond community to work toward a resolution.

Seventeen

MABEL EUGENE

Margie Richard, the center of this struggle for three decades now, had to retreat to the sidelines in the latter half of 2001 to focus attention on mother Mabel.

Since the day three years earlier when Shell officials rushed to the Norco Elementary School, overwhelmed by an over-heating five-thousand-gallon tank of MEK that threatened to repeat the events of 1988, Mabel's health had suffered. Her weakened body, nourished through breathing machines and hooked to a shopping cart of equipment just to get from one room to another, required Margie's daily care. Mabel Eugene suffered a severe bout of asthma, and her frailty now limited her impassioned church speechwriting.

Though Margie remained on the board of Concerned Citizens of Norco and continued to march and lobby for Diamond, she was forced to step aside as president. She turned to her friend, Vice President Gaynel Johnson, to take the mantle. "Gaynel, you got to."

Gaynel prayed on it and then told Margie, "We ain't giving up."

Margie's mission, for now, was to grant Mother's wish: "Whatever you do, I want you to get me away from this plant."

Margie turned to Gaynel again. "My mom has told me three times to get her out but I don't want to leave without you. You stuck with me through the whole thing. When everybody was giving up, you never gave up. You stuck by my side."

"Don't worry about me; get your mom out. Your mom is a sickly person," Gaynel said. Then she delivered her own beam of hope, telling Margie she had found an apartment to rent in LaPlace, miles from the plant.

For Margie, it was another sign it was time to get Mother away. She set about selling the Eugene family's two adjoining Washington Street lots to Shell, her mobile home netting $70,000, and her parents' larger property a small six-figure number.

Once the paperwork was finalized, in October 2001, Daughter laboriously loaded Mabel into her car, oxygen tank and all, and drove away from the plant to the nearby town of Destrehan, and a three-bedroom house with a double garage, wall-to-wall carpeting, a garden plot in back, and more space than their old peach and white clapboard house in Diamond. When Mabel saw the pink roses in the garden, she knew it was home. Mother and Daughter pooled their money to buy the $150,000 house, and on that first day, they sat together and listened to the afternoon breeze and the birds. With the windows pulled open, the daytime air refreshed them.

Stepping inside the redbrick house, Mabel took a seat and soaked the surroundings in. "This is it," she told Margie, reminding her daughter of her dream to one day live in a redbrick house of her own. "I thought Brother was going to get it for me, but you did. This is it."

Mother called relatives on the phone that October afternoon, and she boasted of the pretty interior and the green grass out front. "Thanksgiving dinner is going to be here this year," she told her nieces, cousins, and friends. "You all have to come and see my house."

On that Thanksgiving of 2001, just a month after they bought the dream house in Destrehan, more than sixty people filed in, the men huddled in the living room watching football, the women mingling in the kitchen or in the garage area, and every room of the house filled with voices, laughter, and food. Mabel beamed, surrounded by her grandchildren and great-grandchildren.

Less than a week after the family sat down to its holiday bounty, the state DEQ finalized another proposed penalty against Shell Chemical. The report detailed several years' worth of lapses—citing Shell's failure to conduct monitoring as required, for instance, along with an unauthorized discharge of 345 pounds of 50 percent sodium hydroxide onto the ground and then into a nearby canal. Simply breathing sodium hydroxide can damage the lungs, and high exposures can lead to medical emergencies. Shell, once again, was told to clean up its act.

Yet even with Rolfes and Diamond residents plotting their next move to challenge Shell Chemical, Margie's attention stayed with Mabel Eugene.

Mother's health continued to decline, and shortly after New Year's 2002, she was on the edge.

Margie was not ready to let go. She leaned in against Mabel, her right hand on Mother's cheek, her left hand cradling the back of her head, the daughter now comforting the mother.

It was Mabel Eugene's deathbed, and the two women found comfort and peace together in their last words. Mabel wore her green shiny bed dress, and Margie was wearing a powder-blue and white dress. Caprice, Margie's oldest and grandchild to Mabel, was there.

"You've done a lot of good things for Norco, and people may not understand," Mabel told Margie. "Turn me loose in your heart. You're going to make it."

"Mama, I don't want you to leave me. Naomi's gone. Daddy's gone. Now you."

Mabel told Margie to think of her children, and to care for her aging uncle George, Theodore's brother.

Margie cried, but Mabel could only smile. "You're going to make it, because you're just like me. You've got faith. You're stronger than you think you are. Keep studying that Book."

Mother's cheek was soft, and Margie pressed against it.

"I love you so much," Mabel told her. "Turn me loose."

Mother had false teeth then, and Margie had always been reluctant to even touch them when Mabel asked for help. This day, Margie grabbed the false teeth and put them in Mother's mouth. "I'm not afraid anymore," she said.

"Something's going to happen to you, and it's good. But you won't know until I'm gone," Mabel Eugene said to her only living child.

Margie put a towel to her mother, wiping the sweat.

"This is it. You finally turned me loose. Now, don't you sell this house. Take care of those roses. I want you to get some rest now."

Margie would soon learn what Mother meant about rest. Mabel had arranged her own funeral and picked her own casket in the days before without Margie knowing. She took care of her estate business, knowing that her daughter would need the rest once she was gone. She even wrote her own obituary so that, when she passed, all granddaughter Caprice had to do was fill in the date. She dispatched Caprice to the funeral parlor to pick out just the right casket—"Yeah, that's the one," Mabel said when Caprice showed her a picture of it. She specified her favorite song and scripture, and she had her finest clothes sent to the cleaner.

Margie rubbed the soft cheek some more, and now the mother and daughter both knew the time had come. Mabel opened her eyes, sweated, and took a final breath, her face now at peace on January 28, 2002.

Mabel was laid to rest at St. Rose's Cemetery on River Road in St. Rose. At the funeral were Margie, her two girls, their children, and Naomi's husband and three children, all grown, and hundreds of friends and mourners.

Mother had loved to dress up, so she'd worn a fancy white suit with diamonds, a special hat, and shoes for Mother's Day the year before. The same outfit, freshly cleaned, covered her as she lay in the casket, and Margie wore her mother's hat.

Margie stood up and sang before the eulogy "To God Be the Glory." She felt like she was herself on a cloud, and her voice rose to heights not reached before.

When the singing was ended, Margie's friends and family came to her, and amid embraces and tears they told her they had never heard her voice so beautiful as on the day she said good-bye to her mother, Mabel Eugene.

Eighteen

CLEARING OUT
FOUR DEAD ENDS

At exactly 12 P.M. Thursday, May 16, 2002, the Concerned Citizens of Norco arrived at Shell's U.S. headquarters in Houston. On the same day that Shell held its shareholder meeting at international headquarters in London, the community dramatized its quest for relocation in a demonstration right outside the towering One Shell Plaza.

The fifty-story high-rise was built one year before the Shell complex in downtown New Orleans, and like its sister structure, it exuded wealth and corporate muscle, with an exclusive private club on the forty-ninth floor and a lobby museum for visitors to savor the history of Shell Oil. Outside this grand edifice of 1.6 million square feet, Anne Rolfes and the people of Norco hoisted banners saying "Let My People Go!" So soon after Mabel Eugene's funeral, Margie did not join the trip. Her neighbors carried the message more than three hundred miles, bringing with them a letter from U.S. Representative Maxine Waters, blown up in large size, telling Shell in no uncertain terms to relocate the community. "Shell needs to stop playing games with the lives of families, children, and the elderly," said resident Delwyn Smith, who had taken the reins as president of the Concerned Citizens of Norco after Gaynel had held the position. "We want out now!"

This small group's protest was inspired enough to attract the attention of the local press in Shell's U.S. homestead. "It ended up in the *Houston Chronicle*," Rolfes pointed out. "They were furious."

In Louisiana, Shell officials felt the pressure. Their corporate chiefs wanted to know why they couldn't settle their own mess in the Bayou State, and why they, at headquarters in Texas, were now on the receiving end of this attention.

Yet Shell was giving no quarter on the issue of community health problems. It would never acknowledge that its plant was making residents sick. To support its cause Shell's Brignac handed community lawyer Harden a batch of reports including one from 1997 titled "Illness Absence at an Oil Refinery and Petrochemical Plant," based on the research of the Houston-based Shell Medical Department. In meetings with Diamond activists, the company now made the case that buying out all four streets would actually harm Norco, by hampering the small grocers and vendors who depended on the people in Diamond for commerce.

Harden could only shake her head. Shell insisted that the community buy into the research done by its own doctors, and told Diamond residents what was good for them. "No one voted Shell into any governance office, but they have no hesitation in assuming that role," she said. "'We know what's best for you in this little town.' That's not your call. That was the first hurdle to get over. The next hurdle was how to communicate and talk with them."

Shell and Diamond leaders concluded that only a professional facilitator could help solve the impasse, and they decided to share the cost fifty-fifty to ensure impartiality in the process. Harden told the residents that they needed a facilitator, all right, someone seasoned enough to bring even Israelis and Palestinians together. That's precisely what they got. One of the two local experts eventually brought in had actually worked with Palestinian and Israeli families engaged in neighborhood conflicts. "This is *just* what I prayed for," Harden said.

The facilitators, Jean Handley and Julianna Padgett, were both based in New Orleans. Handley, an American-Israeli, had worked to bridge emotional barriers during her seventeen years in Israel, mediating not only Israeli-Palestinian conflicts but multiple matters where two parties were at odds. Now that she was living in the United States, part of her focus was to bring crime victims together with their assailants to get the criminals to take responsibility for the harm committed.

The facilitators set to work in a church in Diamond, with a half dozen members of Concerned Citizens of Norco and their advocates on one side, and Shell and its corporate lawyers from Houston on the other. In that first session, Handley was struck by the vast divide. "I can remember very clearly meeting in the little room. People were sitting on opposite sides; there was no eye contact," she recalled. "It was really a fragile, delicate balance—a lot of pain, a lot of hurt, a lot of misunderstanding on everybody's part."

She turned to the parties with a simple request. "Put yourself in the other's moccasins," she said.

She and her colleague Padgett had the two sides complete an exercise they surely had never considered before. They were asked to write down on paper all the good things the *other side* had done over the three-decade fight. "What was really important in those earlier meetings," Handley explained, "was to somehow find some ground to build some trust."

The approach might have seemed overly sappy, but the sessions marked a change. The two sides were now talking, face-to-face, and the negotiations offered the prospect of giving the Diamond activists a real voice in resolving the issues, something unthinkable at the dawn of the struggle.

They met for six weeks, and eventually Handley noticed the chill thaw. Residents from Diamond and leaders with Shell soon began greeting each other by name, even as they differed on the details of the battle's resolution.

At the end of one session, the facilitators went around and asked everyone how they felt about how it was going. "Defeated," one said. "Deflated," said another.

It was Anne Rolfes's turn. "Determined," she told the facilitators.

Soon the fight between the four-street community and the international conglomerate reached a national stage, where the entire country could witness what was happening in Norco, Louisiana. A documentary, *Fenceline: A Company Town Divided,* aired nationally on PBS in 2002. In an evenhanded manner, it contrasted life in black Diamond with life in larger white Norco. The documentary could just as easily have been titled *Worlds Apart*.

The long-timers still stuck in their shacks spoke of life next to the chemical plant with a mix of despair and determination, and their children played a game of peekaboo behind signs warning of the plant's chemical dangers.

On the other side of town, where the homes and yards were bigger, one Shell retiree remarked, "It's one of the best companies to work for." David Brignac, a Shell manager, said, "We're catching our samples and saying, 'Uh, no, we see clean air here in Norco.'"

In white Norco, a resident said the plant's ever familiar flaring meant a mishap had been prevented. "Actually, I kind of like to look out at the plant and see all the lights, and if there's a flare going, it brightens everything," the woman said.

Driving through Diamond with the film crew, Margie witnessed the fire blazing orange and turning black. "When that black smoke comes and meets with the dark clouds, you don't know what's a cloud and what's smoke. How can you not help but be in fear?" she asked.

The camera followed Margie as one of her grandsons, Gregory, underwent batteries of tests at Children's Hospital in New Orleans to treat his asthma. "Draw in again. Deep, deep. Blow out. In . . . out," Margie urged Gregory later at home.

In white Norco, Sal Digirolamo told the documentary crew he knew many people in Norco but couldn't think of one person who had asthma. "Not one. And I'm thinking hard," he said.

Next, Wilma Subra provided a guided tour of Cancer Alley. "Only one individual in the Diamond community, only one individual, works at the industrial facilities. The rest of Norco is 98 percent white and these people either work at the refinery or one of the other industrial facilities in the area, or are retired from the Shell facilities."

As the crew filmed, a fire broke out at the Orion Refinery Company sulfur unit, just next to the Shell refinery, and Percy Hollins and Margie Richard rushed out to take an air sample. "I can still smell it. Yeah, that makes me sick. I gotta go," Margie said.

A new housing subdivision was rising in the white section, and Wilma Subra told Shell it made sense to take a blood test on the newcomers before they moved in, and then to monitor their health.

The company declined Subra's suggestion. Yet this latest attention, to be beamed to a national audience on PBS, had hit Shell's image. No, it wouldn't test the health of white Norco. But with the cameras gone, it was thinking seriously about doing something for black Norco, and this time it was an offer that would stick.

Subra was among those sitting across from Shell, or dialing into conference calls, as the discussions careened toward either collision or closure, and she continued to press the company to buy out all four streets. "It makes sense to do this," she told Shell. "These people are living on the fenceline. They are going to be damaged. How can you legitimately let these people be on the fenceline, in harm's way?" she asked.

As the two sides now spoke openly, Subra could see Margie Richard's impact. Years before, when Margie had first spoken with Shell, company officials had been condescending, treating her as just another angry gadfly. As the sides met more often, and Margie spoke in plain language about life in Diamond, about her grandchild with asthma, her sister's death, and her late mother's labored breathing, Subra could see the company finally paying attention.

As Shell weighed whether to move the people of Diamond away from the plant, Subra sensed two palpable, and very different, emotions emanating from the company.

One gave hope that Shell would finally grant Diamond's request.

Shell had actually engaged the Diamond leaders, and Subra saw how the discussions had opened the company's eyes to the community's story. "They were making the effort to listen," Subra realized. "And they were hearing and feeling and seeing what the community felt. And they were hearing it personally." Subra believed strongly that Wayne Pearce, the influential plant manager, wanted to settle in a fair manner.

She also sensed reluctance from Shell to pull the trigger on a full buyout, and she believed it came not so much from Norco as from corporate headquarters. If Shell offered to buy out all four streets of Diamond, wouldn't other communities, across the country and perhaps across the globe, insist on similar treatment? "Shell was scared to death of the precedent. It wasn't the money. It was the precedent," Subra surmised.

One day she sensed a full offer was imminent; the next day she worried that Shell was withdrawing completely. "I said, 'You mean you're pulling out?' And they said, 'We'll have to talk about it and get back to you.' They kept wanting *not* to do it, but still sitting at the table."

Then the company took action.

> <

In a historic reversal on June 11, 2002, just a month before the documentary was to air and weeks after the Houston demonstration, Shell offered to buy the entire Diamond community—Washington, Cathy, Diamond, and East streets. The neighborhood, the family, would not be divided.

The company now offered $80,000 minimum for an owner-occupied house and $50,000 for a mobile home, far more than it had been paying before. Those who opted not to sell could pocket a no-interest home-improvement loan of up to $25,000 that would be forgiven if they stayed more than five years.

"The community is like an extended family and we realize now that our previous efforts to create a greenbelt around our facilities may have created difficulties for some families," Shell managers wrote to Diamond homeowners. Some residents were instantly skeptical, stunned that Shell had actually granted their wish, given in, something that hadn't happened in their lifetimes.

This offer was real.

Shell Chemical's new Diamond Options Program included a booklet sent to homeowners detailing the fine print. Owner-occupied conventional homes would now receive not only the minimum price of $80,000 but also moving allowances and expenses and other bonuses worth potentially thousands more. Vacant lots would get a minimum of $17,500.

Shell's days of paying bottom dollar for Diamond properties were now officially over. A few people on the far two streets hesitated, but most agreed to sell to Shell.

After some of the offers started coming through, the

bitterness of a too-long battle was finally sweetened with tears and shrieks of joy. The Norco activists who had greeted Shell with chilly stares at the start of the formal sit-down talks just weeks earlier now invited company employees to a neighborhood barbecue. The ice had been broken, though not entirely.

L'Observateur, "Best Along the River Since 1913," opined on June 19, 2002, under the headline "New Hope for Diamond": "Now comes the harder part—the healing. Charges of environmental racism have scarred relations, not only between Diamond residents and Shell but between those residents and their other Norco neighbors, and it will take time for those wounds to heal."

Maxine Waters returned to town, and standing in church, she spoke of the power of a community's voice. "Amen," rang out from the crowd.

At Greater Good Hope Baptist Church on Cathy Street, Waters reached out for Caprice, Margie's oldest, the rare soul from Diamond who'd been employed by the plant throughout the years-long struggle. Caprice had been hired by Shell Chemical, but when the company sold a piece of its business to another firm, Caprice joined that company: she was still working at the plant, but now getting a paycheck from a different corporate entity.

"Where is the daughter that works in the plant?" Representative Waters asked. When Caprice rose, the congresswoman asked her to step to the front, where Waters told her how special she must be to hold a job at the same plant her mother had been challenging.

Normally Margie would beam with pride as her eldest was

adorned with praise. But watching her daughter singled out struck a different chord in her. It was a reminder that her flesh and blood worked at the plant that had been her target, and Margie felt her child was exposed. You know what? They're going to get my child. They can't get me, Margie thought. They were only fears, and she put them aside at last, turning instead to the larger quest that had reached its conclusion.

Margie had known this day would come, that all four streets would be included in Shell's buyout, and she was thankful it arrived during her lifetime.

"I want to cry tears," Margie said at a celebration one month later at the Greater Good Hope Baptist Church. "Tears of joy."

One hundred and fifty residents, activists, and Shell employees came together that day, passing out certificates to honor the occasion and celebrating with gumbo and sweet cake. When Shell's Wayne Pearce stepped up to accept his honor for helping negotiate the resolution, the crowd erupted in cheers. Thirty years of pent-up anger at Shell now turned to an emotional thank-you. The company had, at long last, heard its neighbor.

In the coming months, Diamond began its facelift one lot at a time, beginning with the paperwork. Estates had to be settled and appraisals made, first of the value of each property and then of the cost to tear down the house and clear the land. With those details in order, some residents next set out to salvage pieces of their homes before the heavy machinery set to work. Many preserved the strong, durable cypress wood that lined their walls, saving it for the next house they'd inhabit. Others kept entire rooms intact to be moved across town, salvaged valuable copper wire, or cleaned and preserved the bricks that framed their homes. Trailer owners had it easiest; if their home was fairly new and still in good shape, they could simply tow it to a nearby town.

With each owner's piece of history preserved to his or her liking, the wrecking crews tore down what was left, and bull-dozers came in to clear and level the land. A few holdouts decided not to sell at all, but instead to live out the rest of their lives in Diamond. Within two years, only a scattered few homes stood.

As the majority of residents from Washington, Cathy, Dia-mond, and East streets sold out to Shell and found new homes elsewhere, the community's family tree dissolved into a field of grass.

Nineteen

GOLDMAN

With the relocation battle won, Margie Richard had some personal business to tend to. Her family had never been wealthy, not even close, and even though the payment from the Shell relocation package allowed her to fully buy the house in Destrehan, she had to use some of that money to pay debts she had incurred for her daughters' education, to repay loans she had taken over the years, and to cover medical expenses for her family. Margie had always lived on a shoestring budget, and now in Destrehan, the bills continued to mount. She had to pay more in taxes than she'd ever paid in Norco, medical bills still piled up, and cost-of-living expenses only increased.

Margie's father had always preached that his family would not ask for handouts. When someone needed a financial lift, they'd turn to family. With her parents and her sister dead, Margie was the head of the family, and there was no one to turn to in order to make ends meet. Living on the modest paycheck of a retired schoolteacher, Margie crunched the numbers, and the fear hit her that she'd have to sell the redbrick house with the lush garden out back, breaking the vow made on her mother's deathbed. Margie thought back to Mabel's joy those precious few months they lived together in the quiet community. Destrehan is near industry, as is most every place in this corner of southern Louisiana, but it isn't smothered by it like Diamond. The lawns are green and large, and the houses range from handsome to grand. It was home, and now Margie prepared to let it go.

In 2004, her phone rang, the call coming from San Francisco.

At the age of sixty-two, Margie Richard was awarded the Goldman Environmental Prize for grassroots activism, becoming the first black person in North America to receive the prestigious honor. She joined the likes of Wangari Maathai, an earlier Goldman recipient, who would later receive the Nobel Peace Prize for her dedication to Kenya's environment.

The prize brings $125,000 for each winner.

Margie would not have to sell Mabel's dream home.

"I remember standing in my backyard thinking, Lord, will there ever be hope? But a little voice within me kept saying, 'If we don't tell them, how will they know?'" Margie told the Goldman officials after learning of her honor.

On an April day in 2004, Margie and the other Goldman winners were celebrated in a formal ceremony in a San Francisco opera house, with three thousand people in attendance and Robert Redford's voice narrating the stories of each winner, tales spanning the globe that were displayed on a giant screen for the crowd below.

Margie's story came first, and after the segment closed, she stepped to the stage, wearing a formal black evening dress, heart-shaped necklace, and glittering earrings. Her voice forceful, she spoke of environmental racism, climate change, waste disposal, and the energy crisis. Mostly, she spoke of people treating people with dignity.

"The earth is the foundation of life. Environmental issues should never be ignored," she told the San Francisco audience, standing solo on the opera house's stage. "To monopolize the earth for selfish greed or gain is against humanity. Love has no race, no color, no nationality, and no creed." Before she closed, she sang "To God Be the Glory," the song she had sung at her mother's funeral.

The Goldman people put her journey in perspective, writing of how Margie Richard was just one of many minority residents living across from industry in little-known pockets of the country, where pollution chokes the air and residents are more likely to be sickened. "Community protest against these conditions has produced a uniquely American brand of activism that is equal parts civil rights and environmentalism. Richard stands at the forefront of this important social justice movement," the organization wrote, honoring her not just for her work in Diamond but for pushing for change in other fenceline communities as well.

After the Norco buyout, Margie had traveled to Port Arthur, Texas, a poor, black enclave that borders a massive oil refinery and has the highest rate of respiratory illness in the state. Activists say more than twenty thousand of its children are exposed to toxins that can cause cancer and birth defects, and Shell/Motiva is a frequent fixture on grassroots organizers' list of the "Dirty Dozen Industries." Residents along the fenceline in Port Arthur are surrounded by industry on two sides and a canal where ships load and unload on a third. As in Diamond, their daily lives are enmeshed in smoke, flares, and pollution.

Margie journeyed to Port Arthur with a contingent of activists and organizers, driving in a near beeline across Louisiana and into Texas for the nearly five-hour, 240-mile trip. She met with community leaders, sat across the table from government officials, and walked through the neighborhood known as "Gasoline Alley."

It was like stepping back in time to Norco. The people were angry, as they had been in her neighborhood, but had yet to take the fight fully to those they blamed for their plight, industry and government.

"If you don't stand up for something, you will fall for anything," Margie told parishioners inside a community church, mixing her sermon with song. "God gives you the strength you need. . . . You have to take facts into your own hands. It's your battle."

In 2003, within a year of Diamond's resolution, Margie, bucket pioneer Denny Larson, and other activists traveled to the Rubbertown industrial hub in Louisville, Kentucky. Taking Norco's cue, the people of Rubbertown posted signs in their yards demanding buyouts and cleaner air. "I just want out of here," one resident told a reporter for the *Louisville Courier-Journal*. The man, whose home stood near a chemical plant, complained of bright lights, noise, pollution, and hydro-chloric acid vapors that had been released into the air earlier that year.

Wilma Subra, working in Rubbertown for several years, concluded that the community's air was more toxic than Diamond's, as the Kentucky community stood near a lengthier roster of chemical companies, power plants, and rail yards.

A church activist had initially invited Subra to Rubbertown with a simple plea. "Everyone is sick here," he told Subra. "We have a problem and we need your help." When the chemist first ventured there, she made plans to hold a workshop with the community, industry, and local government agencies. Industry and government agreed to meet, but only on their own, not with the community. Subra ended up hosting three workshops.

By the time Margie came to town the divide was closing, but not entirely. Riding in a sightseeing van one afternoon, Margie began choking and sneezing, the air so foul she had to put her coat over her mouth to breathe. As others stepped outside the van to walk in the community, Margie stayed put. "I can't handle this anymore," she said.

Later she took to the church, telling the residents living along the fenceline about her own town's journey. When her speech finished, Margie was flooded with questions. How did Norco advance from complaining about pollution to actually getting Shell to do something about it? How did industry respond when the Diamond activists demanded face-to-face meetings? If you call government, and they ignore you, what do you do? What if people in your own neighborhood disagree with you?

A young person wanted to know what damage the plant pollution could bring to his health. As Margie answered each question and fielded the next, she sensed a resignation in some of the voices of Kentucky, as if the battle had already been lost.

"People were real frustrated," said Subra. "She was there encouraging them it could be done. Relocation was possible."

Years later, the residents in Kentucky and Texas could both point to concrete steps forward. In Rubbertown, some residents along the fenceline have been relocated, some chemical emissions have been reduced, and an air-monitoring program is in place for all to see online.

In Texas in 2006, Shell filed a permit to more than double the size of the Motiva facility, making it the largest refinery in the United States. The refinery, a joint venture with Saudi Aramco, had a capacity to produce 285,000 barrels a day, and Shell envisioned more than doubling capacity to 600,000. Port Arthur activists led by Hilton Kelley kept up the pressure, and later in 2006, Shell/Motiva signed formal papers agreeing to earmark $3.5 million for a community fund that would help residents and local businesses revitalize the neighborhood. Motiva Enterprises also vowed to fund air-improvement programs, including delivering twenty hazard-warning radios to churches and other public buildings, the local press reported. In turn, the residents agreed to withdraw their official protest of the $7 billion expansion.

After Port Arthur and Rubbertown, Margie traveled to grade schools in her native Louisiana and to college campuses from Vermont to Pennsylvania to California preaching environmental justice.

In Louisiana, she and her neighbors hoped their buyout had changed Shell. Even after the settlement, state investigators continued to take note of problems at the chemical plant. In 2002, one month after the Diamond community rejoiced, the state DEQ moved to levy a fine against Shell Chemical for a series of pollution lapses, including numerous plant emissions that exceeded state air-quality regulations and equipment that lacked the required controls. In the course of the inquiry, Shell had been forced to submit a revised air permit application "correcting inaccurate emissions estimates that resulted in the permit exceedances." Shell promised to tighten controls to avoid another layer of fines.

In June of 2004, the Shell Chemical plant and neighboring Shell/Motiva refinery were recognized by state and federal agencies for curtailing environmental problems, welcome news after years of whistle-blowers, inspections, fines, settlements, and nonprofit reports asserting otherwise. Shell said it was monitoring the air and water with more frequency and had reduced air emissions by 30 percent. The company welcomed its first manager to reside in the town of Norco. Not all problems had been erased. Shell reported that in 2004 and 2005 it did not meet its goals regarding flaring from the plant.

Margie Richard was featured in Shell's annual report for 2004. Despite her fight with Shell, if she feels the company— or any other industrial entity, for that matter—makes strides, she stands with it. "The door for communication which was once closed is now open. The right people have been at the table listening to the community's concerns and needs. Trust is now being built," Margie's passage read. "Disagreement is okay because the results can always solve existing problems for industry and fenceline residents."

She concluded, "What happened in Norco between industry and fenceline people and the community as a whole could be a role model for all other facilities because together we can make a difference."

Lily Galland hoped the company's struggle with Diamond had changed the corporate culture. "A lot of it has to do with lack of communication," Galland said from her office off River Road in Norco. "And when I talk with people wherever in the Shell chain, I stress how important it is to communicate."

Across town, the people of Norco remain angry that their beloved company and cherished community had drawn such scrutiny, and some still think it was wrong for Shell simply to buy out Diamond.

Sal Digirolamo thinks the postscript to this struggle speaks to what the community is made of. "We have conquered our differences and now we are all one. If the rest of the world knew you can fight and still be friends, that is a good thing," he said. "The number one thing now is they want to be good neighbors."

Yet Digirolamo speaks with unsettled feelings about the buyout itself. "They are buying out a part of our history, a part of our community," he said. "Diamond was always a piece of Norco, and we just about lost it now."

By the summer of 2007, Diamond's 269-lot community had been whittled to just 9 homes, the final hangers-on in a neighborhood left mostly now to memory.

Well after the Goldman ceremony in San Francisco, Margie walked through Diamond anew, and as she glanced toward the last remaining homes and peered through the fence where the chemical plant hummed, her face betrayed both satisfaction and sadness: joy that her family and neighbors are free to choose where they live, pain that the struggle was coursed with so many losses and could easily have gone astray.

Some memories remained vivid, and real. One of her father's pecan trees still stood, as did his fig, tangerine, and orange trees. Sometimes Margie would drive by Diamond and see a Shell worker enjoying one of her father's fruits, and she herself was never shy to gather a bounty for her own family.

This afternoon, as she walked across the spot where her father's chickens and pigs once roamed, her thoughts spun back through the long journey she had traveled, from the shotgun shack in Belltown to the brand-new Shell Chemical plant that prompted her family to resettle in Diamond to the community fight she had led for more than a decade. Uncle Brother and Aunt Mabel and Naomi were with her now, and for a moment Margie felt as though she were back inside 26 Washington Street.

"It was a quick flash to all the goodness that happens when you stand up for justice. And all of the trials and all of the ills, it's worth it all. I thought about the smallness of Concerned Citizens of Norco."

Ask Margie about grassroots activism, and she'll reach down and grab a few blades from the land where Theodore used to plant his butterbeans and okra. It comes from the ground up, she says, and seeing her childhood street now barren, her friends and family removed from the plant's smoke, smell, and flares, that is proof enough.

In June 2007, Margie Eugene Richard formally asked St. Charles Parish to rename the road nearest the chemical plant Theodore Eugene Street. Her daddy would have been too proud to ask.

AFTERWORD

Margie and her family survived Hurricane Katrina, the devastating storm that seemed to target the most vulnerable and brutally disfigured the historic face of New Orleans. They evacuated to northern Louisiana, where they were housed in a campground run by the Assembly of God. Christopher, Margie's oldest grandson, had attended camp there before, and he negotiated a spot for the entire family just as Katrina's immense winds whipped their way toward the exposed Gulf Coast region stretching from lower Mississippi to lower Louisiana.

The family had typically stayed rooted during hurricanes, but this one was different. Ericker, Margie's youngest, was concerned because her son Gregory, an asthmatic, had to have electrical power to fuel his breathing machine. With Katrina's looming winds sure to carry enough force to knock out power,

Ericker called older sister Caprice. "Will you please come with me to make sure I get there?" Soon, the entire family was en route upstate.

Margie was among the last to leave. It took her more than ten hours to reach the campsite by car, and she finally arrived after two in the morning, but arrive she did, and when Katrina delivered its relentless wallop to New Orleans, her family had settled in a safe harbor. At camp Chris served as night watchman while other members of the family organized dinners and social events.

Margie's Destrehan house took hits to the fence, windows, gazebo, shingles, and trees, but in context that was a small dustup compared with the suffering of the residents of the impoverished city neighborhoods of New Orleans. As the city that birthed her was now buried under water, death, and fear, she saw what everyone else did as the images beamed from New Orleans: the faces of those suffering were almost always black.

Margie heard how the government was offering ready money to those in need, and she thought back to how Shell had lured panic-stricken residents in 1988 with offers of a quick $1,000 if they only signed the legal form. People under pressure, she thought, don't read the fine print. "The people who needed the most were going to be left out," she said as Katrina's devastation became clear. She couldn't let go of the thought.

In May 2006, well after the storm had blown through and at a time the big city was still trying to piece itself back together whole, so many of its residents still housed in hotels, motels, and cars, Margie took a phone call from daughter Caprice, the lab technician working at the chemical plant.

Caprice was calling from her job, and Margie could hear tears behind her words.

Oh God, they fired her, she knew.

Deep into her mother's activism, in 2001, Caprice had been transferred to another company, which had bought a piece of Shell and opened shop at the plant along River Road. Caprice and others working for Shell simply shifted over to a new company letterhead; their jobs remained the same. Still, in the course of the transfer, Caprice's benefits package was curtailed. Margie worried that there was a correlation to her own activism against the plant, but she found solace in the fact that Caprice was still working. "Mom, if you can do anything, try to get me back to Shell," Caprice told her mother one day.

A few years later, there had been yet another corporate letterhead—the second company selling out to a third, which now employed Caprice. That was the firm that was firing her several months into 2006, and its employee roster included some, like Caprice, who had long worked for Shell but then shifted over to the new employer still based at the Norco plant.

With the headlines of Margie's fight faded, and with the Diamond community almost completely a memory, Margie took the call she had dreaded.

"I'm coming right now."

Caprice brought her sobs under control, and she told her mother she would handle the situation herself. The company wanted to escort Caprice and her belongings out the front door in plain view of her colleagues, but Margie's daughter avoided this public spectacle, telling her bosses she didn't need her belongings now, she'd come back for them later. Caprice kept her head on straight, just as Margie had a decade earlier when a principal suddenly was all over her case in the classroom. When she came back to get her box of supplies, Caprice shook hands and thanked the company for the fifteen years.

Shell officials maintain there is no correlation between Margie's activism and her daughter's ouster from the plant. "I don't see how," said Lily Galland, Shell's external affairs manager, who has come to know Margie well. Shell isn't related to the second or third company, Galland said, so how could it be involved in forcing Margie's daughter out? Case closed, the company said.

Margie and Caprice are not so sure, noting Shell's closeness to the new company. They also know this is not a matter that can be proven in black and white, so they moved forward, just as Margie had when she left the school a decade earlier.

Margie collected her thoughts that May morning, and she stepped outside her house in the suburb of Destrehan, where she listened to the birds sing and checked on the colorful flower beds she had taken to planting. Uncle Brother and Aunt Mabel were gone, Naomi was gone, and so many memories of Diamond were too.

"I have to be strong. That's all I know what to do. This morning when I went in the backyard and heard the birds singing and saw my flowers, I thought, This too will pass. We're over the hurt. We're survivors. I must be there for her like my sister was there for me."

Caprice again put her mother's mind at ease. Three months after her dismissal, she made clear she had embraced her mother's words on the day of her firing, when Margie told her: "No weapon forced against you will prosper." She later found work helping supervise cleanups of hazmat, or hazardous materials, sites, and continued to seek employment in a chemical lab.

Caprice had never sugarcoated life for her own two children. Not after her husband, Allen, died, and not now. She'd never been a millionaire, and losing the plant job wouldn't change that. There was still food on the table. "The world isn't always fair," she told her boys, "but you will survive."

Her message came at a joyous, crossroads time for her oldest son, Christopher. She lost her job one week before the family was to take a trip to Arizona for Christopher's graduation from Bible school.

Christopher, eldest of Margie's five grandsons—along with his brother, Joshua, and cousins Gregory, Bradford, and Landon—still calls Margie "Maw-maw," a variation on Grandma, and she looks upon the towering young man with beaming eyes. Christopher was born not long after Naomi's death, and his arrival refocused Margie on what she had, not what she had lost, yet as a child Christopher would suffer loss of his own, his father, Allen Torregano, dying of a kidney ailment when the boy was seven.

This child had long acted like a man, and though he was big enough to play high-school football, a menacing vision as lineman in jersey number 75, and bright enough to choose any vocation, Christopher turned toward religion, as his grandmother had. In eleventh grade he declared he would attend Bible school, and he did.

Two years earlier, in September 2004, Margie had put on a bright purple dress and arrived at the First Assembly of God in nearby Metairie, a building whose plain exterior gave no hint of the passion and sermonizing inside its walls. Every Sunday is special for her, but this one was even more so. Chris was being baptized, and Margie beamed as if it were her own day. As Christopher's head and shoulders washed back in the water, Margie opened her Bible and looked upon the well-worn pages. "For I will defend the city to save it . . . ," read the scripture in Second Kings, nineteenth chapter, the thirty-fourth verse; and in Margie's own hand in the margin were the words "God gave me for Norco." She looked back at Chris and shook her tambourine. Her now-towering grandson was heading off to Bible college in Arizona. By his very early twenties, he had traveled to inner cities and foreign countries spreading the word, always taking care to check in with "Maw-maw."

Now, in 2006, the family traveled to Phoenix to pay witness as Christopher graduated from the Master Commission Bible Studies College in a ceremony at Phoenix First Assembly of God Church. Chris sang solo, and when it came time for a special ceremony—the graduating student literally passing the mantle to a special person, who then passes it back—he called upon his mother, Caprice.

The weekend would be filled with overflowing batches of Margie's crawfish étouffée and potato salad, and her daughters brought more pounds of shrimp po'boys than you could count, much less eat, but Margie put off thoughts of that celebration for a spell as she sat in the vast church in the desert.

"Whenever someone in our family goes through the difficult times, there's always someone else there to give you strength. I thought of how my sister was there for me. For every struggle, for every heartache, there is a silver lining and that was my silver lining. I watched a young man with a strong mind. He will not bend his principles for others.

"Wow, four more to go."

In Ocala, Florida, in the early days of 2006, Margie's lesson was absorbed by a group of community activists who gathered for Bucket Brigade training at the Gospel Temple Church of God and Christ. The people of Bunche Heights, a quiet working-class community in Ocala's black side, had for years complained of ripe pollution and breathing problems they believed were triggered by the Royal Oak charcoal plant next door, a behemoth that baked wood chips into charcoal and whose twin smokestacks churned a sooty smoke into the air. Some of their young children needed nebulizers to treat their asthma, and some of the elderly required oxygen tanks to breathe.

"We have to keep our windows closed 24/7. It's piles of dirt that get stuck inside your screen," Ruth Reed, president of the Neighborhood Citizens of Northwest Ocala, told the people around her in the church that Friday evening in mid-February, and her neighbors nodded. Reed complained that she gets dehydrated often, requiring her to bring a bottle of water to bed each night inside her two-story home just a block from the plant. "Bringing groceries in, the air is so bad you have to run inside."

For years Royal Oak—and Florida's government regulators paid to oversee it—had said all was well, that the air was not foul, and that the sicknesses next door could not possibly be linked to the plant, a refrain familiar to those in Diamond.

Now the people were arming themselves with proof otherwise, and their story felt like an echo of Norco's. Ruth Reed, a woman with drive, speaks with directness, not unlike Margie, and like Margie she is a retired schoolteacher. Sitting inside the quaint church that evening, it was no stretch to think of the parallels to the meetings in the Diamond homesteads two decades before, just after the explosion of 1988. Royal Oak had been built in 1972, a year before the chemical plant eruption killed two Diamond residents.

On this evening the residents of Ocala welcomed a first-time visitor to Florida, the Bucket Brigade's Denny Larson, and they received a primer on the buckets used to test the air. Larson said the community had already taken an air sample from Ruth's house by swiping the crud off cars with a cotton swab and testing it for specific toxins. "You're breathing at least twenty different chemicals, some of which cause cancer," he told the group. "At least we know that now."

Six months earlier, the Ocala community had petitioned the city council to take notice of what was in the air, and the political pressure they brought led state regulators to take a deeper look at the plant. When they did, they found that Royal Oak had emitted illegal amounts of the toxin methanol. Royal Oak, the nation's second-largest charcoal briquette manufacturer, responded to this unwelcome attention: it would close the plant.

Larson said the company's response was all the more reason to test the air, as the plant continued to operate for a short while before shuttering for good.

The next morning, Saturday, everyone gathered again at a community center for more training and hands-on bucket testing. In a small room with a projector up front and a handful of residents fanning around, Larson said that "the design of all these plants is to blow the crap over the fence" and into neighboring yards, and he then delivered a glimpse of what other communities had done to challenge big industry, featuring Norco, Louisiana.

Not long after the weekend brainstorming, with residents energized to test the air for themselves, a company crane went to work dismantling Royal Oak's twin smokestacks one by one, and keeping the company's record of pollution a secret.

In New Iberia, chemist Wilma Subra encountered her own postscript after the Diamond buyout plan had faded from the limelight.

For decades she had operated her Subra Company from a trailer, but in 2000 she moved the headquarters to a house near her home, and there she put the Subra Company sign on display for all to see. Everyone in town knows Wilma Subra, and through her decades of work digging for environmental truths, she has gained admirers and enemies. You start poking into the environmental record of big industry, you're sure to affect somebody's bottom line.

Subra had dealt with the occasional break-in of her workplace, typically as she was scouring the records of some corporation or another. Subra heads the local emergency planning commission, so she knows area sheriff's deputies well, and each time there would be a break-in, a deputy would give her a knowing look. "You know this is harassment. You know this isn't a normal burglary," the sheriff's deputy confided.

At 7 P.M. on June 14, 2006, a slow-moving car crept in front of the office as Subra worked inside with the light on. A shot rang out, and the bullet pierced the building's front brick exterior. Her husband, working on the flower bed at their home around the corner, called the sheriff's deputy. The culprit escaped.

"The first thing I started doing was going back and seeing what I was doing that was controversial. The previous times I was broken into, I could link it back to an issue," Wilma explained. At the moment, she had a handful of controversial issues spread out on her desk. No telling which battle she was then fighting had triggered this shot, but Subra believed it had to be one of them.

She was not hit, but the blast made an impression. Wilma Subra, MacArthur Foundation fellow, the chemist with the disposition of a grandmother and the persistence of a cop, installed bulletproof glass in her office's front windows, and she moved her computer station to the back, not so visible to passersby.

Then she went back to work along Louisiana's Chemical Corridor.

Though the Diamond community is now just a few homes scattered on the historic grounds, big industry churns all around it.

In 2005, the Shell Chemical Company boasted a payroll of more than $50 million in Norco, with 546 employees and another 163 under contract, a big-budget reminder of its economic impact on St. Charles Parish.

In the summer of 2006, St. John the Baptist civic leaders beamed with the word that more than $2 billion in projects were under way or planned at three massive plants—Marathon Petroleum's refinery near Garyville, a sugar refinery west of Reserve, and a DuPont plant in LaPlace.

Company officials talked of the jobs that would come, both in construction in building the plants and on-site once the expansions were complete, and local politicians obliged with nodding approval, singing industry's praises as employer, taxpayer, neighbor. One local leader in St. John the Baptist Parish had worked at Marathon for twenty years, and it was clear where her sentiments lay.

"Parish permits are unlikely to pose any problems," the *Times-Picayune* noted, rather matter-of-factly, in its account of this growth on the river.

In St. Charles Parish, other expansion plans were afoot, and again the economic development officials talked of the jobs that would come. Dow was building a new vinyl methyl ether plant, a coup for the local decision makers who beat back other communities scrambling for this piece of commerce involving a product that helps make water treatment sanitizers. Near Diamond, Valero's one-thousand-acre refinery, which already processed a quarter million barrels of oil each day, was laying the groundwork for its own major expansion that could nearly double its capacity. Homeowners witnessed the work crews dismantling old equipment to make way for new, and they knew big industry would continue to grow.

With the flares and the clouds come jobs, taxes, and profits. In southern Louisiana, crude's reign surely continued. Big Oil vowed its growth would not come at a price to the air or the community, and when the industry talked of the millions it would spend to keep emissions clean, local politicians took the companies at their word.

One afternoon in January 2007, the industry of Norco flexed its muscles with particular intensity, the smoke-filled air visible for miles away on Airline Highway as it churned upward in giant puffs. Even on a gray, overcast day, the chalky plant-produced clouds stood out from the real ones above. In southern Louisiana, no one seemed to notice.

ABOUT THE
RESEARCH

*N*ight Fire is a true story span-
ning five decades, built upon public records, private documents, and the recollections of the people living in or drawn
to the Diamond neighborhood of Norco, Louisiana.

The manuscript is equal parts oral history and recorded
documents.

The oral history comes in part through more than thirty
interviews with Margie Richard. The first came in September
2004, when we walked through Diamond as Margie recounted
life in the shadow of industry and the day in 1973 that she
witnessed two neighbors killed in the streets. A day later, I sat
with Margie in the First Assembly of God Church in nearby
Metairie during her grandson Christopher's baptism. Since
that first weekend meeting we spoke many times, often while

sitting in Margie's kitchen or living room in the suburb of Destrehan, but also over meals and through multiple telephone conversations. Margie's story drives the narrative from beginning to end, and the quotes printed here come primarily, though not exclusively, from those interviews. Other quotes were taken from public statements Margie made over the years in a lawsuit deposition, in community forums, during government meetings, in letters, and to documentary film and international radio crews. These verbatim transcripts were obtained during the research, and in each case in the text, I have tried to make clear when Margie's comments were made in such public forums. Otherwise, the comments were made in interviews with the author or, in a few cases, in the court deposition involving her suit against Shell.

When Margie's comments are in direct quotes, it reflects dialogue that she and other interview subjects helped re-create, or quotes that come from verbatim transcripts of the court deposition, media interviews, and governmental meetings.

The story of the community and the chemical plant was enhanced by many other voices, and they included interviews with residents from Diamond and the other side of Norco, Shell Chemical officials, Margie's family, community activists, lawyers, experts, scientists, and government regulators.

Thousands of pages of documents were gathered in researching these events. Most notable among them was the six-volume court archive of *Richards v. Shell*, the community's lawsuit that went in the company's favor in 1997. Other vital files include the Louisiana Department of Environmental Quality (DEQ) inspection and citation reports on Shell's neighboring chemical plant and refinery; filings and reports by officials for the federal government and agencies such as the Environmental Protection Agency (EPA) and Occupational Safety and Health Administration (OSHA); and brochures and literature written by both community activists and Shell Chemical.

What follows is a chapter-by-chapter description of the key research elements gathered in the preparation of this book.

Chapter 1: DEATH IN DIAMOND

Over the course of any story, there are crucial moments that give the narrative shape, context, and heft. The Diamond community's struggle with Shell Chemical produced many such important dates: the time Shell bought out property in Norco in the mid-1950s to build its new chemical plant; the explosion at that plant two decades later that left two residents dead; the death of Margie's sister, Naomi, a decade later; the even more violent eruption at the Shell refinery in 1988; and so forth.

During my visits to southern Louisiana, I gained a flavor of the Chemical Corridor while driving along Airline Highway and River Road, and by sitting for interviews with multiple sources.

Yet to gain a richer feel for life in this region, particularly going back decades, I can think of no better exercise than the hours spent one morning in March 2006 at a grand three-story building in LaPlace, Louisiana—the St. John the Baptist Parish Library. There I read decade after decade of *L'Observateur*, a local newspaper, printing out grainy black-and-white pages of the paper's news coverage before, during, and after many of the key events happening nearby in Norco. The details from this small newspaper helped me set the scene in certain sections of the manuscript, but nowhere more importantly than in chapter 1. I learned about the school savings deals in a full-page ad in the newspaper, about the Sugar Queens who were anointed and the crime spree that arrived, about the strawberries in bloom, the expansion of major industrial plants along River Road, and the impact the gas shortage of 1974 had on the communities of this region. When I sat down to write about important events in Diamond, particularly during the 1950s and 1970s, these pages helped bring the region alive with the small details of life in neighborhoods far from the glare of New Orleans.

Local news coverage can help only so much. The crucial event of 1973 in Diamond, the summer explosion at Shell Chemical, generated extremely limited news coverage. I was able to track down two stories in the archives of the New Orleans *Times-Picayune* about this explosion—a modest eighteen-paragraph page 1 piece the day after the accident, and a short brief a day later. No other news stories could be found.

Likewise, public records about this tragedy proved elusive. I asked Shell Chemical about this event, and the company said it had no records and only limited recollections to share. I filed Freedom of Information Act requests with multiple government agencies—OSHA, the Louisiana DEQ, and the Norco Area Volunteer Fire Department—yet these requests generated zero public records of relevance.

Researching the deaths of Helen Washington and Leroy Jones, I felt as if I were chasing ghosts. Their deaths have largely been wiped off Louisiana's public records.

To learn more about the events of that summer day in 1973,

I interviewed Margie Richard in Diamond; she recounted where she saw Helen Washington dead under a white sheet and how she glimpsed Leroy Jones trying to outrun flames after his lawn mower was lit in the explosion. I interviewed several other longtime Diamond residents who witnessed this tragedy, including Margie's cousin Doris Pollard, and spoke with activist Anne Rolfes, who, like Margie, had interviewed the mother of the teenage boy killed that day. These oral histories helped shape the narrative of the summer day that changed Diamond.

I learned details about the Eugene family through public records and multiple interviews with Margie Richard and her family. Court records provided the date and price of the Diamond homestead Theodore Eugene bought in 1954, and brochures compiled by community activists and Shell provided information about the plant's expansion and operations. For details about Margie's father's fight to open a high school in his neighborhood, I was aided by a June 1992 anniversary piece in the *St. Charles Herald* that included lengthy comments from Theodore Eugene.

Louisiana State Department of Education records detailed Margie's graduation from Grambling College, and news archives, Wikipedia, and Web sites devoted to historical information provided details about the slave revolt of 1811.

For the larger, richer story of Margie's parents, I was aided by family snapshots, video footage taken for a PBS documentary, an interview with Margie's daughter Caprice Torregano, and, most important, Margie's recollection of growing up with Uncle Brother and Aunt Mabel.

Chapter 2: SISTERS AND SECRETS

This chapter was built largely upon a series of lengthy interviews with Margie about her sister, Naomi. Margie's recounting of their closeness, both as sisters and fellow teachers, shaped the narrative, as did her recounting of the illness that took her sister's life. I read medical literature from the U.S. Department of Health and Human Services to learn more about the disease that afflicted Naomi, sarcoidosis.

I interviewed Margie's daughter Caprice at length about her perspective on the racially tinged school shooting at Destrehan High School, where she was a student.

Details on the trailer that Margie made her home in Diamond were gathered through court files from the community lawsuit. Telling details about the struggles in southern Louisiana of the 1970s—the racial unrest, the gas shortage—came through interviews and newspaper archives from multiple media outlets.

Chapter 3: PEARL HARBOR

Margie's personal recollection of this explosion at Shell's refinery across town aided my reconstruction of the events of that early morning. Other interviews with area residents—with Diamond's Gaynel Johnson, with Shell employee and activist Sal Digirolamo, and with Shell executive Lily Galland—provided additional details from the viewpoints of their respective homes.

Public records and news archives fleshed out this chapter in important ways.

A U.S. Department of Labor OSHA Accident Investigation Summary provided numerous details about the explosion at 3:37 A.M. on May 5, 1988, including the cause of the eruption, the extent of injuries inside the plant, and facts about the seven fatalities that morning.

DEQ inspection reports were the source of material for the two citations noted in this chapter that occurred prior to the deadly 1988 blast. I obtained those records, and many other such public documents, from the DEQ's office in Mandeville, Louisiana, during August 2006 while exploring enforcement actions brought against Shell's twin plants since the 1970s. For a description of the hazards of ethylene, focus of the 1988 DEQ citation, I was aided by a to-the-point New Jersey Department of Health hazardous substance fact sheet.

I closely read coverage in the New Orleans *Times-Picayune* about the 1988 tragedy, and the text benefited from the ground-level reporting provided by that newspaper's thorough coverage. Also perused was coverage in *L'Observateur,* the Associated Press, and the *Baton Rouge Morning Advocate.* This coverage, for instance, included details about how the temporary shuttering of this one plant affected consumers nationwide.

Chapter 4: **A MISSION**

To learn about the beginnings of Margie's grassroots campaign, I interviewed Margie for hours, but also spoke at length with several neighbors who were on the ground floor during this community uprising, including Gaynel Johnson and Janice Darensbourg.

The community's subsequent lawsuit against Shell, filed in 1993 and tried in 1997, also provided crucial details—such as the fact that the Norco Relocation Committee's initial dues were $2 or $5 per person.

Chemist Wilma Subra, who had worked with the Diamond community since the 1970s, shared her recollections of the early struggle and provided informative reports about pollution and health hazards in this region as we drove through Diamond and larger Norco during a visit in February 2006 and in subsequent interviews.

Public records were the source of quotes involving a Shell hazardous waste permit application, as DEQ files included the verbatim minutes of community public hearings concerning this application in August 1989.

A United Press International wire service dispatch detailed the Citizens Fund toxic pollution report that cited Shell and Norco.

Chapter 5: GRASS ROOTS AND LAWSUITS

Interviews with Margie and her daughter Caprice, the lab technician hired at the Shell plant, form an important part of this chapter, but numerous other interviews and documents were helpful. The Louisiana Bucket Brigade filed a report years later showing how much Shell had paid for land in Diamond, and that report is cited here and elsewhere in the manuscript.

Lengthy interviews with Patrick Pendley, the Plaquemine lawyer who helped file the community's lawsuit against Shell in 1993, provided on-the-ground information about the legal strategy adopted in the community's case against Shell. I met Pendley at his office in February 2006, and there the lawyer shared his experience working not only on the Diamond case but on similar fenceline fights in Louisiana. Pendley later fielded some of my follow-up calls about his legal team's strategy in Norco, which came into question as Shell handily won the case. Pendley was not able to provide a clear answer as to why his legal team declined to call chemist Wilma Subra or researchers from Xavier University to the stand, other than to say the team presented what it believed was a winning case. I sought interviews with his co-counsel in the community lawsuit, Allen Myles, but Myles did not respond.

This chapter includes other information gleaned from Margie Richard and the community's lawsuit, notably the initial twenty-five-page complaint filed in 1993.

Chapter 6: "A FAST ROUTINE"

Interviews with Margie provided information regarding the 1988 explosion and, later, Shell's one-stop shop where residents could collect $1,000 and sign away the right to sue. I interviewed Shell official Lily Galland in Norco about this legal offer, which the company defended as fair and, to the residents who accepted it, generous. I asked Galland about Margie's description of her meeting with a Shell lawyer at the company's New Orleans office. "I can't say either way with what happened behind closed doors," Galland said. "I know legal things can be confusing at some times, but I stand behind Shell."

Newspaper archives in the *Times-Picayune* were helpful in

explaining how the giant settlement over the 1988 explosion was handled in the legal system.

I gathered profile information about Shell's corporate office in New Orleans, at One Shell Square, and those details helped set the scene regarding Margie's meeting with a lawyer there in 1993 to discuss the larger explosion settlement. Similar profile information was gathered regarding Shell's U.S. corporate headquarters, One Shell Plaza in Houston, which is described later in the text.

Chapter 7: *RICHARDS V. SHELL*

Few chapters benefited more from public records than this. I spent hours one day reading and copying large parts of the six-volume court file of *Richards v. Shell* at the St. Charles Parish Courthouse, and the details in the file brought the community and company's struggle to light, each side's case spelled out in black and white.

The file included the initial complaint, of course, but also other instructive files, including Margie Richard's 112-page deposition in 1994. That deposition included the disclosure that Shell started calling Margie only *after* the lawsuit was filed, and that the calls had gone to her employer, the public schools, not her home. Not much later, the public schools started to come down on Margie, and St. Charles Parish school records later detailed Margie's retirement after thirty years in the system. I tried on four occasions to seek comment from the principal who pushed for Margie's ouster while the lawsuit was moving in the courts, but the principal, now working at another educational institution in southern Louisiana, did not respond. I interviewed Shell's Galland about Margie's ouster from the schools, and though Galland was not directly involved in the lawsuit, she said she understood how Margie could surmise that the school crackdown was a result of her case against Shell; Galland said, however, that she doesn't believe the company had a role in Margie's teaching status or subsequent exit from the schools.

Galland was generous with her time, meeting in person, responding to follow-up questions, and providing information ranging from a video detailing Shell's history in Norco to reports about the company's operations. I sought an interview with the Shell community point person at the time, Don Baker. He declined an invitation to share his perspective on these events or to respond to concerns from Margie and some of her neighbors that he, as Shell's liaison with the community, had paid little heed to their views. "He's not interested in talking to you. He retired and he's moving on in his life," Galland explained. Other Shell figures did not respond to requests to share their perspectives; among them was corporate counsel Charles Raymond, who was heavily involved in the company's legal strategy over the years, including its successful defense of Diamond's civil court lawsuit.

From Margie's personal files at home, I obtained letters schoolchildren had written to her after the death of her son-in-law, which occurred around the time the principal was seeking to fire her.

The *Richards v. Shell* court file included legal descriptions of many of the homes in Diamond, including their date of purchase, sale price, and site size, all details that enriched the text. It also included reports filled out by residents complaining of sickness, depositions of other Diamond property owners, and consultant expert reports filed by both plaintiff and defendant that provided a clear window into the focus of their respective cases as *Richards v. Shell* headed to trial.

I interviewed Pendley, lawyer for the community, and Galland, Shell's external affairs manager, about the respective approaches to this legal battle. An interview with Norco civic leader Sal Digirolamo helped shed light on the other side of town's view of the Diamond community's legal challenge of Shell.

From Margie Richard, I obtained copies of flyers and other grassroots literature she and her neighbors created during their quest to get Shell to relocate them. I also interviewed Samuel Coleman, an EPA enforcement official based in Dallas, who had visited Diamond after hearing Margie speak in 1995.

Chapter 8: PROOF

This chapter is largely built from the "Community Health Survey: Norco/Old Diamond Plantation," which was submitted to Xavier University's Deep South Center for Environmental Justice in 1997. The "Old Diamond News," another Xavier University document, included interviews with Diamond residents, and some of those sentiments are cited here.

The chapter also includes information gleaned from court records in the community's suit against Shell, including dueling court filings filed by both sides, and from pollution reports filed by chemist Wilma Subra. It also includes an interview with activist Gaynel Johnson.

Chapter 9: IN THE COURTROOM

This chapter benefited largely from files obtained in the case of *Richards v. Shell.* Among its thousands of records, the court file included exhibit lists filed by plaintiff and defendant as the lawsuit went to trial, providing a fingerprint of each side's case, and it included some handwritten notes taken by the court reporter detailing arguments in court. The file also included motions from both sides spelling out the focus of their cases, along with references to hundreds of state DEQ reports involving Norco's air quality.

The court file included psychiatric evaluations filed by a Shell expert who had examined several of the core plaintiffs in Diamond, and sections of those evaluation reports are cited here. Ultimately, the court file included the jury verdict itself, ten to two in favor of Shell.

I interviewed Margie at length about the case and her reaction to it, interviewed her neighbor Gaynel Johnson and chemist Subra, and read *Times-Picayune* trial coverage.

Chapter 10: THEODORE EUGENE

This narrative is built from a lengthy interview with Margie about her father's final days, and benefited from family snapshots taken of Theodore Eugene during the holidays not long before his death, plus an interview with Margie's daughter Caprice.

Chapter 11: "MINOR INCONVENIENCES"

This is another chapter built, in part, from court files, including a postverdict filing by Shell's lawyers that likened life in Diamond to no more than a series of "minor inconveniences."

Other public records were important, including a DEQ inspection report detailing hazardous waste concerns at the chemical plant, this one in 1997.

Shell was facing scrutiny outside of Louisiana, and I obtained government filings regarding a September 1998 lawsuit settlement involving the company in Illinois, including comments from then attorney general Janet Reno and others.

An Amnesty International report produced in 1994 is the source of information on the questions over the conviction of Gary Tyler. Also crucial to this chapter was the Shell/Loyola University survey of Norco community opinion about industry, which was obtained and is quoted at length.

Interviews help shape this section, including one with La-Place attorney Randal Gaines, who represented Margie and her neighbors in a subsequent pollution scare involving a Mother's Day lime spill. Gaines shared court files from previous civil rights cases he had handled in Louisiana. Literature compiled by community activists included accounts of the Mother's Day lime spill.

I also interviewed Caprice, Margie's oldest daughter, about the company's reaction to the lawsuit that was shared with her inside the plant not long after the jury verdict; interviewed Diamond's Gaynel Johnson; and spoke with Earthjustice lawyer Monique Harden about her introduction to both Margie and the Diamond community.

Shell provided written statements to the author about the lime spill and tank overpressurization.

Chapter 12: NORCO ELEMENTARY

For this chapter, I obtained a brochure, put out by Shell and St. Charles Parish, regarding what to do in event of a chemical emergency, and quoted it directly here.

I also obtained detailed government and other reports to describe the events of December 8, 1998, when two mishaps put a sharp focus on life in the Diamond community.

Among the records: an "Air Sample Collected Report" written by Wilma Subra after the community's air was tested that day, and a report by Loyola University's Center for Environmental Communications that included important facts about the "unusual events" of December 8, 1998. Also obtained was the U.S. EPA's formal complaint against Shell Chemical Company involving the events of that afternoon, which contended that Shell had violated the Clean Air Act; and an executive summary of the NEJAC conference that had been held in Baton Rouge at the time.

Multiple interviews about this day's events and significance were conducted, with Margie Richard, Bucket Brigade leader Denny Larson, other residents of Diamond, Wilma Subra, lawyer Monique Harden, and Shell officials. In addition, I spoke with EPA official Samuel Coleman, who attended the Baton Rouge conference, and who is now regional director of the EPA's Superfund Division.

Also informative were two *Times-Picayune* stories involving December 8, 1998—one the day after the scare and, more significantly, a takeout by writer John M. Biers in September 2000, headlined "Bad Air Day," that included a blow-by-blow recounting of this day.

Around this time, Margie Richard was beginning to make the rounds of environmental groups asking for help for Diamond, and transcripts of several of those speeches were obtained: to an international radio program, to the UN Commission on Human Rights, and to an EPA conference.

DEQ reports are the source of the January 1999 fine for the release of chloride in the community, and for the summary of such citations against Shell over the years. A New Jersey Department of Health hazardous substance fact sheet provided details of the health hazards of allyl chloride and MEK.

A *Times-Picayune* story was helpful in detailing some of the bucket testing in Norco, and the research benefited from multiple media reports concerning the work Erin Brockovich and Ed Masry had done in California.

Shell provided a written statement to the author concerning the events of December 8, 1998.

Chapter 13: ALLIES

Extensive interviews were conducted with key allies drawn to Diamond's struggle with Shell, including Anne Rolfes, founder of the Louisiana Bucket Brigade. Rolfes shared reports she and other activists compiled in the battle with Shell. I obtained the "Shell-Norco Toxic Neighbor" report, compiled by the Sierra Club and other groups, that was the basis of the community's "November surprise" upon Shell. Also quoted is a report Rolfes had written involving another battle against Shell, in Nigeria. A *New York Times* column by Bob Herbert provided insightful details about Shell's issues in Nigeria.

Community activist reports and multiple interviews helped round out this chapter, most notably detailed discussions with Margie and Anne Rolfes about the day they each, separately, interviewed Ruth Jones, whose son died in the 1973 blast.

Other interviews were conducted with Maura Wood of the Sierra Club, who helped Margie place the "FlareCam" outside her trailer in 1999; with Wilma Subra; and with Commonweal founder Michael Lerner.

From Margie, I obtained a printed page from the community's Web site detailing the chemical plant's flare burn on one such day in July 1999. Margie and her neighbors began keeping diaries when the air was foul, and several of those dispatches are cited. I also obtained a speech Margie made before an EPA group in Virginia, and it is quoted verbatim.

This chapter benefited from coverage of the community's struggle with Shell in the *Times-Picayune*, including reporting of an Environmental Defense Fund pollution report that cited Shell Norco.

Chapter 14: THE WHISTLE-BLOWER

DEQ documents regarding the joint state-federal settlement with Shell/Motiva provided details, as did the U.S. Department of Justice press release outlining the case. Noteworthy records were obtained at every step of this investigation, from the initial DEQ statement detailing the case and its potential penalties, to the final settlement between DEQ and Motiva, to the consent decree between the parties. All of these documents helped flesh out the story of the investigation spurred by the Shell Oil refinery whistle-blower, Barry McCormick.

Separately, the whistle-blower's civil complaint against Shell detailed the allegations he had brought against the company. I interviewed the whistle-blower's lawyer in New Orleans, and this chapter also benefited from enterprising coverage of the whistle-blower's case and Shell's mounting pollution troubles by John M. Biers, then the energy reporter for the *Times-Picayune*.

Chapter 15: THE OFFER

I obtained letters Margie wrote to Shell after receiving the offer she believed would cut the neighborhood in half, and quoted from a brochure compiled by the Louisiana Bucket Brigade decrying the idea of dividing Diamond.

Shell shared its own reports spelling out the company's buyout offer to the community, and I interviewed Shell's Galland at length about these negotiations and, later, the ultimate resolution between company and neighborhood.

I interviewed a Norco community leader about his neighborhood's view of Shell's resolution, obtained more air quality reports from Wilma Subra, and gathered records involving the joint state-federal settlement with Shell/Motiva and others, a settlement that was occurring near the time of the company's offer to the Diamond residents.

Newspaper coverage of the battle between Diamond and

Shell helped provide a time line of key events. For verbatim quotes on the day Margie presented a bag of Norco air to a Shell official, I benefited from a transcript and footage from the documentary *Fenceline: A Company Town Divided,* which captured the scene. The film was directed by Slawomir Grunberg.

Chapter 16: "TOXIC PARK"

Key to this chapter were letters exchanged by Congresswoman Maxine Waters, Margie Richard, and Shell's CEO in Houston. Also informative were copies of protest letters from Diamond residents and a dozen members of the U.S. Congress that were sent to Shell.

Detailed reports from the Louisiana Bucket Brigade and Shell brochures provided helpful information. Also important were now more constant local newspaper reports about the discord, and further reports from Wilma Subra.

Also obtained in the research was a September 2001 study concerning pollution and health problems in Louisiana from Environmental Health Perspectives; and, separately, an Associated Press report about a deadly explosion at a Motiva facility in Delaware.

New Jersey Department of Health hazardous substance fact sheets provided information detailing potential problems associated with benzene and sulfur dioxide.

Chapter 17: MABEL EUGENE

This chapter was built almost entirely from a lengthy interview with Margie Richard about her mother's final three months in life, beginning with the day they moved into their new house in Destrehan and closing with her mother's funeral.

It also included an interview with daughter Caprice, who was also at Mabel Eugene's deathbed, and with Diamond activist Gaynel Johnson.

DEQ documents were the source of the citation at the chemical plant involving sodium hydroxide, and, again, a New Jersey Department of Health hazardous substance fact sheet provided details of the hazards of this corrosive chemical.

Chapter 18: CLEARING OUT FOUR DEAD ENDS

Interviews with community activists, residents, and Shell officials helped me to describe the ultimate resolution, including one with facilitator Jean Handley, who worked to help bring the two sides together. Also informative was the documentary *Fenceline: A Company Town Divided*, which told the nation about the two sides of Norco, Louisiana. I obtained the documentary and its script. Wilma Subra spoke of Margie's impact on the fenceline fight with Shell, and of the frantic final negotiations between the two sides in the weeks leading to the final offer.

The text benefited from writer Tina Susman's first-rate story in *Newsday* on July 14, 2002, which provided details of the community and company's celebration one month after the buyout. Subra described for me, in vivid detail, the process of Diamond owners preserving pieces of their homes before moving elsewhere.

Chapter 19: GOLDMAN

I interviewed Margie Richard at length about the community settlement and her receiving the Goldman Environmental Prize, and obtained documents and videos from Goldman officials in San Francisco that told why Margie was chosen for this honor and how she was celebrated in the ceremony.

A story in the *Louisville Courier-Journal* provided telling information about yet another fenceline fight, this one in the Rubbertown area of Kentucky, and news reports and interviews with activists were the source of information about Port Arthur, Texas.

Interviews with Margie, Subra, and others fleshed out this chapter with details about the battles of other fenceline communities.

State DEQ files and Shell literature were the sources of information about the Norco plant's problems and progress.

In Louisiana, I walked with Margie as she recounted growing up in the home of Uncle Brother and Aunt Mabel, the grounds now covered mostly in green grass.

AFTERWORD

Interviews with Margie and her daughter Caprice were vital, particularly as they relate to Caprice's ouster from her job as a lab technician at the chemical plant. I also spoke with Shell's Galland about this turn of events, and sought comment from the company (not Shell) that fired her, but did not hear back.

One of the earlier talks with Margie came just as Hurricane Katrina was heading toward southern Louisiana. At the time, Margie wasn't sure whether she would stay in Destrehan or make the long drive up to northern Louisiana, where her family was staying. I spoke with my wife, who was in our home in South Florida, where hurricanes have wreaked serious havoc, and called Margie back with a plea that she evacuate. She did, and ten hours later arrived to join her family, far from Katrina's most savage blows. Some two weeks later, while out of town conducting research for an unrelated newspaper series for the *Miami Herald*, I heard back from Margie that she was okay.

Also aiding this closing section was a reporting trip in early 2006 to Ocala, Florida, where Denny Larson was introducing his Bucket Brigade concept to another minority community residing within shouting distance of industry. I spent a couple days observing this grassroots training and interviewing Larson about Norco. Diamond's struggle, and resolution, served as a case study for this community in northern Florida. I was on leave from the *Miami Herald* at the time, working on this book, but the story in Ocala was so pressing that I passed along information about it to a gifted colleague, Cara Buckley, who wrote a powerful series, "Air of Suspicion," that helped inform my research. While in Ocala, I met Hilton Kelley, an activist leading another fenceline fight, in Port Arthur, Texas; he had come to Florida to help Larson introduce the Bucket Brigade to Ocala residents.

In Louisiana, over lunch of shrimp po'boys purchased at a favorite haunt in Destrehan, I interviewed Wilma Subra about the bullet that pierced her Subra Company offices.

I obtained Shell literature detailing the chemical plant's

budget and staffing numbers. The *Times-Picayune* in 2006 wrote a story about expansion along the Chemical Corridor, and some of those details are described here. Among my visits to Louisiana was a January 2007 trip during which I observed Norco's billowing industry clouds while driving along Airline Highway.

ACKNOWLEDGMENTS

This book came together with the support and generosity of many people.

At HarperCollins Publishers/Amistad, I am grateful that editorial director Dawn L. Davis saw potential in my proposal and believed it was a book. The manuscript was enriched by the graceful line edits provided by Dawn and her colleague Christina Morgan. It also benefited from the diligent copyediting eye of Adam Goldberger.

At the Zachary Shuster Harmsworth literary agency, Esmond Harmsworth and Mary Beth Chappell answered all of my questions with sage advice and unyielding support. Their insights helped me better shape this narrative. I also am indebted to Nicholas Maier, who read a story I wrote in the *Miami Herald* in 2004 and dispatched an e-mail I cherish.

At the *Miami Herald*, my wife and colleague, Beth Reinhard, was my biggest supporter and sharpest-eyed critic, and she never failed to find time in her nonstop schedule to take our girls out so I could quietly work. Hearing Abby's and Emma's voices return to the house hours later was a sweet conclusion to a day's work. I owe thanks also to Carl Hiaasen, who took time to read my proposal and offer meaningful support.

Sandy Reinhard, my father-in-law, and his lawyer colleague and friend, Manny Garcia, reviewed the slew of contracts a first-time author signs and generously shared feedback.

In Louisiana and beyond, many people took time to discuss these events, and most later answered multiple follow-up queries. In particular, I thank Margie Richard, Caprice Torregano, Anne Rolfes, Denny Larson, Lily Galland, Monique Harden, Gaynel Johnson, Doris Pollard, Sal Digirolamo, Randal Gaines, Patrick Pendley, Jay Alan Ginsberg, Janice Darensbourg, Maura Wood, Jean Handley, Samuel Coleman, Michael Lerner, and Ruth Reed. My research was aided by material provided by helpful staff members at Xavier University's Deep South Center for Environmental Justice, the New Orleans *Times-Picayune* library, and the Goldman Environmental Prize in San Francisco.

I owe a unique thanks to chemist Wilma Subra, who drove hours from her home in New Iberia to meet me several times in Norco, and then provided guided tours. Subra answered innumerable follow-up questions about everything from the type of chemicals discharged from the plant to the type of wood that lined homes in Diamond. Late in the process I felt it important to send a working draft of my manuscript to an expert to review for accuracy. Subra was the obvious choice. Of course she said yes.